Lecture Notes in Computer Science 2497
Edited by G. Goos, J. Hartmanis, and J. van Leeuwen

Springer
Berlin
Heidelberg
New York
Barcelona
Hong Kong
London
Milan
Paris
Tokyo

Enrico Gregori   Giuseppe Anastasi
Stefano Basagni (Eds.)

Advanced Lectures on Networking

NETWORKING 2002 Tutorials

Springer

Series Editors

Gerhard Goos, Karlsruhe University, Germany
Juris Hartmanis, Cornell University, NY, USA
Jan van Leeuwen, Utrecht University, The Netherlands

Volume Editors

Enrico Gregori
CNR-IIT Institute
Via G. Moruzzi, 1, 56124 Pisa, Italy
E-mail: enrico.gregori@iit.cnr.it

Giuseppe Anastasi
University of Pisa, Department of Information Engineering
Via Diotisalvi 2, 56122 Pisa, Italy
E-mail: g.anastasi@iet.unipi.it

Stefano Basagni
Northeastern University, DANA 312
Department of Electrical and Computer Engineering
360 Huntington Ave., Boston, MA 02115, USA
E-mail: basagni@ece.neu.edu

Cataloging-in-Publication Data applied for

A catalog record for this book is available from the Library of Congress

Bibliographic information published by Die Deutsche Bibliothek
Die Deutsche Bibliothek lists this publication in the Deutsche Nationalbibliographie;
detailed bibliographic data is available in the Internet at <http://dnd.ddb.de>.

CR Subject Classification (1998): C.2, H.3.4-7, H.4, H.5.1

ISSN 0302-9743
ISBN 3-540-00165-4 Springer-Verlag Berlin Heidelberg New York

Springer-Verlag Berlin Heidelberg New York
a member of BertelsmannSpringer Science+Business Media GmbH

http://www.springer.de

©Springer-Verlag Berlin Heidelberg 2002
Printed in Germany

Typesetting: Camera-ready by author, data conversion by PTP-Berlin, Stefan Sossna e.K.
Printed on acid-free paper   SPIN: 10870724   06/3142   5 4 3 2 1 0

Preface

This book of proceedings brings together papers on the topics of the tutorials given at the second IFIP-TC6 networking conference "Networking 2002", held in Pisa, Italy, on May 19–24, 2002.

The aim of this book is to provide overviews and surveys on prevailing topics in the field of networking, to sum up and complement what the speakers conveyed in their tutorials. Despite the vast spectrum of topics in networking, we were pleased to receive proposals from leading researchers in areas of great interest. The papers collected in this book attest to the great variety of topics and the richness of current research and trends in areas that span from mobile computing and wireless networking to network security and optical networks.

The reader unfamiliar with the specific topic of a tutorial will find the papers presented in this book useful as pointers to the basics on each topic. Researchers in the area of a given tutorial will find here the latest trends and achievements on the tutorial topic.

The general chair and the tutorial co-chairs would like to thank here all the tutorial speakers, whose effort and cooperation made this "extra" book possible.

May 2002

Enrico Gregori
Giuseppe Anastasi
Stefano Basagni

Organizers

Sponsoring Institutions

Networking 2002 Organization Committee

Conference Executive Committee

General Chair:
Enrico Gregori, National Research Council, Italy

General Vice-Chair:
Ioannis Stavrakakis, University of Athens, Greece

Technical Program Chair:
Marco Conti, National Research Council, Italy

Special Track Chair for Networking Technologies, Services and Protocols:
Andrew T. Campbell, Columbia University, USA

Special Track Chair for Performance of Computer and Communication Networks:
Moshe Zukerman, University of Melbourne, Australia

Special Track Chair for Mobile and Wireless Communications:
Guy Omidyar, National University of Singapore

Tutorial Program Co-Chairs:
Giuseppe Anastasi, University of Pisa, Italy
Stefano Basagni, Northeastern University, USA

Workshop Chairs:

Workshop 1 — *Web Engineering*
Ludmilla Cherkasova, Hewlett-Packard Labs, USA
Fabio Panzieri, University of Bologna, Italy

Workshop 2 — *Peer-to-Peer Computing*
Gianpaolo Cugola, Politecnico di Milano, Italy
Gian Pietro Picco, Politecnico di Milano, Italy

Workshop 3 — *IP over WDM*
Giancarlo Prati, Scuola Superiore S. Anna, Italy
Piero Castoldi, Scuola Superiore S. Anna, Italy

Invited Speaker Chair:
Fabrizio Davide, PhD, Telecom Italia SpA, Italy

Organization Chair:
Stefano Giordano, University of Pisa, Italy

Publicity Chair:
Silvia Giordano, Federal Inst. of Technology Lausanne (EPFL), Switzerland
Laura Feeney, SICS, Sweden

Steering Committee Chair:
Harry Perros, North Carolina State University, USA

Steering Committee Members:
Augusto Casaca, IST/INESC, Portugal
S.K. Das, The University of Texas at Arlington, USA
Erol Gelenbe, University of Central Florida, USA
Harry Perros, NCSU, USA (Chair)
Guy Pujolle, University of Paris 6, France
Harry Rudin, Switzerland
Jan Slavik, TESTCOM, Czech Republic
Hideaki Takagi, University of Tsukuba, Japan
Samir Thome, ENST, France
Adam Wolisz, TU-Berlin, Germany

Electronic Submission:
Alessandro Urpi, University of Pisa, Italy

Web Designer:
Patrizia Andronico, IAT–CNR, Italy

Local Organizing Committee:
Renzo Beltrame, CNUCE–CNR, Italy
Raffaele Bruno, CNUCE–CNR, Italy
Willy Lapenna, CNUCE–CNR, Italy
Gaia Maselli, CNUCE–CNR, Italy
Renata Bandelloni, CNUCE–CNR, Italy

Table of Contents

Peer to Peer: Peering into the Future 1
  Jon Crowcroft, Ian Pratt

Mobile Computing Middleware 20
  Cecilia Mascolo, Licia Capra, Wolfgang Emmerich

Network Security in the Multicast Framework 59
  Refik Molva, Alain Pannetrat

Categorizing Computing Assets According to Communication Patterns ... 83
  Dieter Gantenbein, Luca Deri

Remarks on Ad Hoc Networking 101
  Stefano Basagni

Communications through Virtual Technologies 124
  Fabrizio Davide, Pierpaolo Loreti, Massimiliano Lunghi,
  Giuseppe Riva, Francesco Vatalaro

A Tutorial on Optical Networks.................................... 155
  George N. Rouskas, Harry G. Perros

Author Index 195

Peer to Peer: Peering into the Future

Jon Crowcroft and Ian Pratt

University of Cambridge Computer Laboratory,
J J Thomson Avenue,
Cambridge CB3 0FD {Jon.Crowcroft,Ian.Pratt}@cl.cam.ac.uk

Abstract. In this paper, we survey recent work on peer-to-peer systems, and venture some opinions about future requirements for research.

The paper is a survey to support the tutorial at the Networks 2002 Conference and is therefore neither complete, nor likely to be up-to-date by the time you are reading this, since the topic was extremely fast-evolving at the time of writing.

Instead, we try to bring some historical perspective and structure to the area, and to shed light on where the novel contributions, and where it is likely that there are research questions to answer.

1 Introduction

The topic of peer-to-peer networking has divided research circles in two: on the one hand there is the traditional distributed computing community, who tend to view the plethora of young technologies as *upstarts with little regard for, or memory of the past* – we will see that there is evidence to support this view in some cases. On the other hand, there is an emergent community of people who regard the interest as a opportunity to revisit the results from the past, with the chance of gaining widespread practical experience with very large scale distributed algorithms.

The term *peer-to-peer* (P2P) came to the fore very publicly with the rise and fall of Napster[1]. Although there are prior systems in this evolutionary phase of distributed computing (e.g. Eternity[45]), we choose to limit the scope of this survey to the period from "Napster 'til Now"[1].

1.1 Background

Peer-to-peer networking has come to refer to a family of technologies and techniques for organising distributed applications, that has emerged over the past four-to-five years.

The consensus seems to be that a peer-to-peer system can be contrasted with the traditional twenty-five or more year old *client-server* systems: client-server systems are asymmetric; the server is distinguished as running over some longer period of time and looking after storage and computational resources

---

[1] i.e.1998-2002

E. Gregori et al. (Eds.): Networking 2002 Tutorials, LNCS 2497, pp. 1–19, 2002.

for some number of clients; the end system is therefore a single bottleneck for performance and reliability. Server sites may make use of a number of techniques to increase reliability and performance, such as replication, load balancing and request routing; at some point along the evolution of this thinking, it is a natural step to include the client's resources in the system – the mutual benefits in a large system, especially given the spare resources of today's clients, are clear.

Thus peer-to-peer systems emerge out of client-server systems by removing the asymmetry in rôles: a client is also a server, and allows access to its resources by other systems.

A claim sometimes made about peer-to-peer systems is that they no longer have *any* distinguished node, and thus are highly fault tolerant and have very good performance and scaling properties. We will see that this claim has some truth to it, although there are plenty of peer-to-peer systems that have some level of distinguished nodes, and also plenty of peer-to-peer systems that have performance limitations. In fact, the fault tolerance claims are hardly born out at all in the early instantiations of the peer-to-peer movement. Initial availability figures in Napster[1], Gnutella[2] and Freenet[3] do not compare favourably with even the most humble of web sites!

However, second and later generation systems do indeed provide the claimed functionality and performance gains, and we will see in Pastry[40], Chord[11] and CAN[12] very promising results, and even more recent work building applications and services over these systems shows great potential gains[16].

At the same time as peer-to-peer, a number of network researchers have been frustrated in their attempts to deliver new network services within the context of traditional telecommunications or Internet networks[2]. Instead, researchers have built experimental infrastructures by constructing *overlay* systems. An overlay may be a simple as a collection of static IP in IP tunnels, or as complex as a full dynamic VPN ("virtual private network"). Some of these systems are in use in the Active Networks research community. Others are more straightforward in their motivation, such as the GRID communities' requirements for more robust Internet connectivity!

The classical distributed systems community would claim that many of these ideas were present in the early work on fault tolerant systems in the 1970s. For example the Xerox Network System's name service, *Grapevine*[43] included many of the same traits[44] as systems mentioned above and discussed below. Other systems that could easily be construed as architecturally true peer-to-peer systems include Net News (NNTP is certainly not client-server) and the Web's Inter-cache protocol, ICP. The Domain Name System also includes Zone Transfers and other mechanisms which are not part of its normal client-server resolver behaviour.

Also, IP itself is built out of a set of autonomous routers running a peer-to-peer protocol (for example, the routing protocols OSPF and BGP are peer-

---

[2] New Internet network level services such as IP QoS in the form of integrated services, and differentiated services, as well as novel service models such as multicast and mobility have proved notoriously hard to build and deploy in their native forms.

to-peer, and certainly not client-server, and are designed that way for exactly the reasons given above for peer-to-peer); not only that, but IP was originally an overlay service , implemented above other layered communications system: the PSTN, ARPANET and X.25 circuit switched networks. Indeed, this overlay model keeps re-emerging as network operators deploy faster switched infrastructures such as Frame Relay, ATM and WDM and PONS (Pure Optical Networked Systems) core networks. Most of the peer-to-peer work that we discuss in this document is positioned at the Application Layer, as illustrated in Figure 1.

One can look at differences and similarities between classical client-server and modern peer-to-peer systems on another axis: statefulness. Despite successes with stateless servers, many Web servers use cookies and other mechanisms (Web Services) to keep state over various transactions with a client. Peer-to-peer systems may keep track of each other although the implications of the privacy/anonymity models we discuss below mitigate this.

Yet another viewpoint from which one can dissect these systems is that of the use of intermediaries. In the Web (and client-server file systems such as NFS and AFS) we use caches to improve average latency and to reduce networking load. Peer-to-peer systems (particularly Freenet[3]) offer intermediary services very naturally, although again, anonymity (and locality) may not necessarily be preserved.

As an aside, we would say that since this topic seems to recur every few years, there appears to be something compelling and unavoidable about peer-to-peer and overlays.

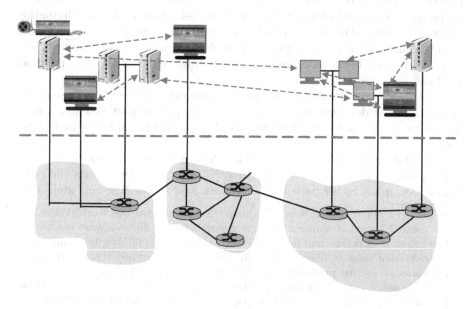

Fig. 1. Peer-to-peer architectural situation

In the next section we take a further look at both this recent, and longer term history in a little more detail.

1.2   History

While we could talk about the 1970s and 1980s work on autonomous distributed systems, or the use of ideas from those in other areas such as IP routing, here we choose mainly to look at the recent peer-to-peer work. [3] Lest readers think *we* are not aware of the longer history of peer-to-peer protocols, we cite the work of 1215[46].

Peer-to-peer "year zero" can effectively be set to Napster[1]. Napster was used heavily to share music content. This is a political hot-potato – the music industry was (and still is at the time of writing) very slow to make content available through the Internet, perhaps due to lack of perception of the market, perhaps due to lack of efficient rights management, and content tracking and protection mechanisms. This seems commercially naïve, given that a large part of the customer base were early network adopters with low tolerance to profiteering. Indeed, the media industry as a whole has preferred to pursue legal mechanisms to protect its legacy hard copy business than to pursue a lower cost (but not necessarily lower margin or profit) soft copy distribution business – the fear of copying would be plausible if it were not for the fact that this risk already exists, so the lack of availability of online music services until fairly recently is all the more remarkable[4].

Napster was not just a system architecture. It also became a company which provided a central directory. This directory is where file listings are uploaded by peers, and where lookups go to. The true behaviour is that of alternating client-server rôles. This is illustrated in Figures 2, 3, 4 and 5.

The central directory server exhibits problems which are not just reliability and performance bottlenecks, but are also a single point of security, political, legal and economic attacks.

To work around all these problems, there has been an evolutionary sequence of research and development of next generation peer-to-peer systems, largely improving on the lookup mechanism, but then moving on to better and better content availability.

The first of these was Gnutella, which dispensed with the central directory and replaced it with a flood based search [2] [35]. While this averts the problems above, it has rather pathological network load effects. The next steps were

---

[3] It has been remarked, we believe by Roger Needham, that the World Wide Web has created a form of amnesia amongst researchers who find it almost too easy to find recent work online, and forget to check the large body of older literature offline in traditional print in libraries. Luckily, this is being rectified through concerted programs at various publishers, notably the ACM and IEEE to scan and make available online entire back catalogs. However, one can see several instances of the lack of historical perspective today in papers which cite nothing before WWW year zero, submitted to recent conferences in which the authors have been involved.

[4] One might observe that the best way to ensure that kids do something is to ban it!

Overlays. As the ease of development and deployment of peer-to-peer became clearer to the network research community, the use of what one might call "multi-hop" applications, or more normally *overlays*, has started to take off in a big way.

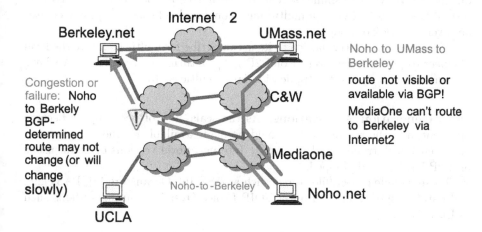

Fig. 6. RON

Balakrishnan *et. al.* at MIT have developed the Resilient Overlay Network system[6,33]. This idea is illustrated in Figure 6, where a number of sites collaborate to find a longer path at the IP "level", which has better properties (such as throughput or loss) at the application level, by dynamically routing via a set of dynamically created tunnels. In parallel, Turner *et. al.* at Washington developed a similar approach for multicast[30].

The difficulties in deployment of native IP multicast have led several groups, most notably the CMU group have developed an "End System only Multicast" system which constructs distribution trees for applications by running a routing algorithm for a degree-constrained multicast tree without including any intermediate IP routers, but with fairly limited negative impact on the efficiency of the distribution tree [7,8]. Further work used this to implement multimedia conferencing systems and demonstrated the workability and usability of the scheme [9]. A group at Stanford also used this approach for streaming media [28].

Other researchers have used the approach for anycast [14], and server selection [15].

Next Generation. Dissatisfied with the poor resilience and network load caused by super-nodes and flooding, respectively, a number of researchers have been working on distributing the directory for peer-to-peer storage systems.

The main approach is to implement a distributed hash table. There are then a number of design choices as to how the keys are distributed amongst nodes, and how requests are routed to the appropriate node(s).

The principle schemes are Pastry[40], Chord[11], CAN[12] and Tapestry[10].

Chord uses a clever *fingertable* to reduce the size and failure probability of the directory. Pastry uses a similar scheme but with network locality hints. CAN uses several hashes, to map into a multi-dimensional node id space. Tapestry creates an inverted index of nodes.

The success of Pastry has led to its use in Scribe for multicast above the level of a peer-to-peer service (overlay on P2P!)[22]. In turn, the creators of CAN have also investigated its use for application layer multicast[13].

Next Generation Applications. As the scalability of peer-to-peer became apparent, and the resources that could be made available on the large number of systems on Intranets and the public Internet, novel versions of applications on P2P have been developed.

These include object/file stores, such as ASF[16], Oceanstore[18], PAST[29], and databases[31], and even novel VOIP (Voice Over IP) routing systems such as Bazooka[32].

Measurement and Security. There is an interesting tension in measurement work in peer-to-peer systems. Due to their open nature, one can introduce instrumented nodes very easily. Due to the anonymity of some of the systems, it can be hard to ascertain the real behaviour of the whole system at a higher level[19].

Some of the more interesting work has centred on the combined behaviour of users and the peer-to-peer system itself. The seminal paper "Free riding on Gnutella"[20] shows that even mutual benefits do not stop people behaving selfishly.

Other aspects of peer-to-peer that have garnered study are security and collective behaviour [23,25].

Modelling Work. Alongside measurement, we usually find modelling, and peer-to-peer networking is no different from telecommunications or the Internet in this regard.

Examples of interesting models include papers on the power law of organisation of the system and the small world model of user, content and peer connectivity [21,26,27,34,36].

Industry and other interest. Peer-to-peer has attracted a lot of industry interest [37,38,39].

Possibly the most interesting work to date is the Microsoft Farsite project[42]. Building a massive, highly available, distributed file system out of (massively) unreliable components obviously has its attractions for this organisation.

2   P2P Taxonomy

In this section, we form a rough taxonomy of peer-to-peer systems. Following that, we take a slightly more detailed look at each aspect of peer-to-peer through some examples.

Examples of peer-to-peer system claimed properties include:

Server-less. As stated above, *true* peer-to-peer systems auto-configure with a minimum of initial state, and then run without any distinguished node. Having said that, many systems build a structure which may include tuning the peering relationships to improve locality [41] and possibly adding hierarchy to improve the performance of directory lookup (size and robustness) and searching. However, this structure is usually maintained through soft-state, and is largely in place in most peer-to-peer systems to improve performance rather than to add specific functionality.

Arbitrary Scaling. Peer-to-peer systems often add resources as they add customers. Thus they should scale (at least at the computing and storage resource level, if not networking) linearly, or better, with number of peers. Of course, many networks are designed assuming client-server traffic, and so it is entirely possible that this performance scaling properties many not be achieved transparently. There are some claims that the "small world" models of human behaviour and interaction, combined with the preferential way in which people link to known locations, lead to power law structures in connectivity. This in turn leads potentially to even better than linear scaling in the system in terms of potential locality (in Freenet for example, however this claim is disputed by some researchers). This means that the study of peer-to-peer systems is partly a sociological, partly an ecological, and partly an economic one.

Symmetry. Software for peer-to-peer systems is symmetric. Such software systems are harder to write than client-server systems at the moment as there are few toolkits (although JXTA[4] is one such system). The peer-to-peer programmer needs to build event driven systems, which in general have proved harder to write correctly than client/server. One risk in such systems is that one has to pay attention to synchronisation effects. Another problem is that the programmer must cope with the type of erroneous requests that only server (rather than client) application programmers have to deal with. This makes peer-to-peer programming currently an expert systems programmer task, rather than the separation of concerns that client-server architecture achieves.

File Sharing. To a large degree, peer-to-peer is used to share content. Napster and its descendants, have been used for media distribution. Gnutella is also being used by the Genome project for distributing the database.

Processor Cycle Sharing. There are few peer-to-peer CPU sharing systems. SETI@Home is one of the few, although (like Napster) it has a super-node. Xenoservers[17] are a system attempting to fill this gap in the *true* peer-to-peer portfolio. SETI@Home is also in use for other applications such as genome database searching and cryptographic cracking.

Anonymity and Resistance to Censorship. Some peer-to-peer systems offer privacy at the level that users' identities are masked. Some go further and mask content so that neither peer exchanging data knows who delivered or stores which content. True anonymity is typically two layer, requiring some form of anonymized IP routing system (e.g. onion routing) as well as application layer mechanisms to protect privacy.

High Availability. Eternity provides a high degree of availability by striping and replicating objects using FEC like codes over multiple nodes. This provides some level of assurance that single (or some level of multiple) node outages, do not lead to object non-availability.

Directories and Super-nodes and Searching. A number of peer-to-peer systems are not strictly peering, but depend on some designated node or server (or cluster of nodes and servers), to head-up the system. Some systems volunteer peers to perform this function. This appears to be hard to avoid if one wants to provide a scalable search or browse capability.

Application Layer Routing. Peer-to-peer routing is application layer. It can be used to overlay IP network layer routing to offer more services such as multi-path, multi-metric, multi-cast, mobile optimisation, and other techniques.

Interestingly, it is hard to find any successful decentralised P2P networks. These are the ones we would say have been most successful so far:

- Napster - centralised
- Gnutella - started out decentralised, but then evolved into a centralised system[37].
- KaZaA/Morpheus/FastTrack - centralised (only super-nodes route queries, central registration/authentication/index server)
- DirectConnect - centralised (pretty similar to FastTrack, except the more you share, the more nodes you get to share with)
- EDonkey - same as FastTrack and DirectConnect (why do people spend time doing the same things as everyone else?)
- Freenet - not successful yet, perhaps because it is decentralised
- MojoNation - also not successful yet, perhaps because it is decentralised

It is interesting to ask what is the problem? Is it lack of bandwidth? Or is it because people don't like to share? Later we look at failure to share[20].

Request and Content Routing. Peer-to-peer systems still have actual *users* who make requests. These requests (for a resource) have to be routed to a node that can provide that resource (storage, object, CPU etc). Request routing has evolved from super-node indexes[1], and flooding[2], through to sophisticated content-key hash based routing[40,12].

Overlays. As we have observed, peer-to-peer systems carry out application layer routing. One parsimonious use of peer-to-peer techniques is *only* to provide an overlay for IP routing to make up for lack of an IP layer built-in service.

3   P2P Detailed Examples

3.1   File Sharing

File sharing was one of the first great successes of client-server distributed systems. Sun RPC and NFS were ubiquitous in the mid 1980s in research and education labs. Nowadays, we can add Samba and other systems. However, the brittleness of such systems has been remarked on, most famously by Lamport – the failure of a so-called stateless server causes systems to "hang" all around a network. Automounting mitigates against this somewhat but without completely eliminate the problem.

Solutions through server replication are typically expensive.

Peer-to-peer file sharing starts out as simply offering directories of files on *my* system to any and all other systems on a network. There are no guarantees that I will keep these files, or that I will serve them to anyone, or what resources I will give to serving a file to someone. Nor, typically, is there a way in early peer-to-peer file sharing applications, to upload a file, only a directory.

Later systems have rectified these shortcomings, but still struggle with notions of performance, availability and persistence.

3.2   Process Cycle Sharing

Processor cycle sharing is familiar through remote shell and earlier in RJE systems. Since Moore's "law" appears still to describe the growth in new CPU speed over time, we have several hundred million personal computing and workstation machines now on the public internet which have hundreds of MIPs going free all the time.

Attempts to exploit this vast processing resource have been limited since the unit of computation is still quite small, but more importantly, since the latency and bandwidth for code and data distribution to these nodes is typically very poor. This means that the class of computation for which it makes sense to use this huge distributed parallel, unreliable computer is rather restricted to jobs of the type that can be broken in to many relatively small pieces which have quite long individual processing time and perform little communication (i.e. the ratio of CPU step to job cycle step is large). Nevertheless problems in large scale signal processing (SETI@Home, Genome searching, Astrogrid, code cracking) have all been successfully tried, and continue to thrive[5].

Recent moves towards finding locality in peer-to-peer systems might lead to the use of clustering, allowing computations of a less constrained style (e.g. in the Computational Fluid Dynamics area, for example in turbine design, atmospheric and oceanographic modelling).

---

[5] SETI is the Search for Extra Terrestrial Intelligence, which takes the signal from many radio telescopes around the world, and attempts to find low entropy strings in it which may or may not represent *artificial* signals from remote intelligent life. Anecdotally, when applied to RF signals sourced from planet Earth, the system failed to find anything remotely intelligent!

3.3   Anonymity and Censor-Proofness

One of the main novel characteristics of Eternity and subsequent peer-to-peer file systems has been the ability to withstand censorship. This is achieved by several means:

Partition. By splitting a file into component parts we make sure that no single site carries the whole file, and a denial of service attack has to run over multiple sites. Later systems made clever use of techniques such as Rabin fingerprinting or other techniques for finding common elements of objects can also ensure that overlapping content between multiple files is exploited to reduce storage costs.

Replication. By replicating blocks of a file over multiple sites, we also provide higher availability. Combined with locality information, this can be tuned to reduce latency and increase file sharing throughput, at some cost in terms of consistency in update.

Encryption. By encrypting blocks of the file, we make sure that disclosure is unlikely, but also that a node can deny knowledge of the actual content it carries – again, P2P exploits mutual benefit: the argument is that "I might not approve of this content, but I approve of the ability to hide my content in a set of peers, so I will live with what I do not know here!". This is often termed "plausible deniability" and is used by Service Providers to align themselves with the "common carrier" defense against legal liability for content, as with telephony and postal providers can do.

Anonymization. By masking the identity of sources and sinks of requests, we can also protect *users* from the potential *censor* or unsavoury *agency*. However we need to mask location information as well as identifiers, as otherwise a traffic analysis may effectively reveal identities, so some form of onion routing is also usually required.

3.4   Directories and Super-Nodes

As has been mentioned, there is a trend for users to wish for searching facilities, including underspecified queries that return a mass of partial matches. To provide this there has been a move to create super-nodes which continually scan peers and construct indexes. Of course these distinguished nodes are now highly visible, and this rather reduces the idea of anonymity for themselves, and possibly partly for the set of peers too.

3.5   Overlays

Overlays are a good technique for implementing services that might otherwise belong in a lower layer. Certainly, at one extreme one could imagine that an idealised *anycast* service should be able to do almost all of the things that peer-to-peer storage systems do. However, the burden of putting this sort of information, even at the level of hints, into the network level addressing and routing infrastructure is clearly too high. Instead, an overlay is created. In some cases, the overlay may in fact serve the purpose of using additional information merely to route packets (i.e. the basic network service, but controlled using more complex application layer hints).

3.6   Request and Content Routing

The notion of a "middle box"[47] or *proxy* pre-dates peer-to-peer by a few years. Proxies use meta-data and state information to decide on routing queries to an appropriate one out of many possible (and hopefully, though not always) identical servers. Many of the query routing systems in peer-to-peer architectures take this idea to a logical extreme.

4   Shortcomings and Future Work

In this section, we look at seven areas where we believe additional research questions can be identified based on problems and shortcomings in peer-to-peer systems today. No doubt other researchers have other topics that they can add to this list, and may also feel that some items on this list are perhaps less important, but these are our current favourites:

Measuring and Modelling P2P Systems. It seems that there is a great opportunity to build interesting systems for the end user with peer-to-peer technology. However, users then have the opportunity to offer traffic from multiple sources simultaneously. This might start to make normal user traffic look a lot like Distributed Denial of Service ("DDoS") attacks!

Update and Persistence in P2P Storage Systems. Most of the classic peer-to-peer storage systems are actually ways to share files, and do not easily permit scalable update, and certainly do not offer Service Level Agreements on persistence.

Computation. We really only have very early experience with sharing CPU resources. The economic motives are not the same: the mutual benefit arguments which work for file (music/film) sharing are not so obvious. Even with Digital Rights Management properly in place, file sharing makes efficient use of link bandwidth (upload while I download can take place on most networks since links are usually roughly symmetric in capacity provisioning).

The economic incentives are not so clear with CPU resource sharing, although the Xenoserver[17] project is one attempt to rectify this with accountability and fine-grained charging.

QoS and Accounting. Peer-to-peer systems do not typically offer QoS (latency or throughput) guarantees. Even if they did, the accounting system could possibly undermine anonymity. Again, the Xenoserver work attempts to tackle this problem.

Locality versus Anonymity. Peer-to-peer systems struggle with the apparently inherent contradiction between offering anonymous sharing of resources, and localisation of service offers. Without a substrate of anonymous packet level routing such as onion routing, it is hard to see how to reconcile this. Having said that, an approach such as recursive use of peer-to-peer, first to build an overlay that achieves anonymized request and response routing, and secondly to do content distribution might be worth exploring.

Evolution of P2P into Lower Layers. As peer-to-peer approaches become better understood, we might see the techniques migrate into the infrastructure, as IP has migrated from overlay into native services.

P2P and Ad-hoc Wireless Network Duality. Peer-to-peer systems incorporate many techniques for self organisation in an ad-hoc scenario without any pre-built infrastructure. It has been observed by quite a few people informally that these characteristics are shared with ad hoc wireless networks. There is perhaps some mileage in exploring this commonality.

4.1   Measuring and Modelling P2P Systems

Large scale Internet measurement has been recommended/proposed by the US National Academy of Science in their report on "looking over the fence at networking research"[48]. In fact they also recommended looking at overlay systems as a general approach to building research infrastructures.

4.2   Update and Persistence in P2P Storage Systems

The Eternity file-system incorporated file block striping and replication in a peer-to-peer system some time ago. This surely needs revisiting, especially in the light of recent advances in the design of redundancy coding algorithms such as those used in layered multicast schemes.

4.3   Computation

The critical difference between file objects and computations is that the former have a simple partition function. Computations are notoriously hard to distribute. There are some interesting cases which are well understood. The various world wide GRID research programmes have a number of such problems which entail very long timescale and large computations, which have some resemblance to the SETI@Home problem.

4.4   Congestion Control versus QoS and Accounting

Recent peer-to-peer systems have a stage during the phase of setting up peering relationships when they use some simple tool such as the ubiquitous "ping" program, to select preferred peers by *proximity*.

However, such an approach is error prone, and more importantly, leads to a succession of local optimisations, rather than a solution optimised for global performance.

It is clear that regardless of whether we want an adaptive system that responds to congestion (e.g. via Explicit Congestion Notification or pricing) or an engineered solution based on some type of signaling, there is a long way to go in providing performance guarantees into P2P. Certainly, a market driven approach would fit the spirit of P2P best. However, as measurement projects produce more results to characterise the behaviour of peer-to-peer sources, it may be possible to develop traffic engineering, and even admission control and scheduling algorithms for the nodes.

4.5   Locality versus Anonymity

As mentioned before, there are a number of driving factors that are reducing the anonymity characteristic of peer-to-peer systems, at the very least, their immunity to traffic analysis is being lost.

However, developing an onion routing system as a peer-to-peer application, and overlaying this with any other peer-to-peer application seems like an elegant, but as yet untried solution to this tension.

4.6   Evolution of P2P into Lower Layers

We observed that successful overlay systems sometimes migrate over time into the infrastructure. However, many peer-to-peer systems are complex, and devising a minimal service enhancement in the lower levels that would support their algorithms is an interesting challenge. The IETF FORCES working group has been working on remote IP router control (separating out forwarding and routing). However, we would need more than this if very general P2P intermediary functions are to be carried out at wire or fibre (or lambda!) speed. Filtering and processing of content keys would be needed. Such a system is a long way from our current capabilities or understanding.

4.7   P2P and Ad Hoc Wireless Network Duality

Peer to peer systems are a highly attractive way to build dis-intermediate content services. However, resilient mechanisms that have emerged in the P2P community from Freenet and Eternity, and via Content Addressable Networks have weird effects on network utilisation and are prohibitive for wireless ad hoc networks. We propose to select some of these mechanisms, such as CAN, Chord, Pastry, and Mixnet, Publius, Xenoservers, and MojoNation and mitigate these effects. This could be achieved by:

1. making them "load aware";
2. distributing load information and implementing a distributed congestion control scheme;
3. adding a topographical metric based on location specification and using anycast.

Most current second generation peer-to-peer overlay systems use content request routing schemes that relay on arbitrary hash functions to map from some attribute (content) to node identifier – in the case of CAN there are as many hash functions as there are dimensions on the hyperspace.

None of these takes account of actual location. These functions are largely aimed at providing robust and anonymous content location, and a high degree of cooperative, distributed indexing. There have been some attempts to modify schemes to take account of actual node proximity to the clients/requesting nodes. Freenet and other systems migrate copies in a manner similar to demand caching, as often used in the World Wide Web.

It is likely that next generation systems will use actual location and scope information to influence routing functions so that content is initially placed and requests are routed to copies that have proximity on a number of QoS axes – these would be delay, as well as potentially throughput and loss-based, but could also include battery budget considerations for wireless users. Thus the distribution of replicas in the service infrastructure would evolve to meet the user demand distribution, optimising use of the scarce wireless resources to better match user concerns.

Note that some of the resulting algorithms and heuristics serve a dual purpose: they can be used for the actual packet routing decisions too! The duality between self-organisation in P2P content location and routing, and ad-hoc wireless routing has been commented on in the past, but not directly exploited to our knowledge.

Acknowledgements. The authors gratefully acknowledge discussion with Jim Kurose and Don Towsley, and the use of materials generated by their class on peer-to-peer from last year in the Computer Science department at the University of Massachusetts at Amherst. Thanks are also due to Ian Brown, Ken Carlberg, Tim Deegan, Austin Donnelly and Richard Mortier for extensive proof reading.

References

1. Napster. "The Technology Behind Napster", copy of this is cached at:
   http://gaia.cs.umass.edu/cs591/hwpa/napster.htm
2. The Gnutella Protocol, version 0.4,
   http://www.clip2.com/GnutellaProtocol04.pdf
3. "Freenet: A Distributed Anonymous Information Storage and Retrieval System", I. Clarke, B. Wiley, O. Sanberg, T. Hong, in Designing Privacy Enhancing Technologies: International Workshop on Design Issues in Anonymity and Unobservability, Springer-Verlag LNCS 2009, ed. by H. Federrat, Springer: New York (2001).
   http://freenet.sourceforge.net/index.php?page=icsi-revised
4. "Sun's Project JXTA: A Technology Overview", L. Gong,
   http://www.jxta.org/project/www/docs/TechOverview.pdf
5. "Morpheus out of the Underworld", K. Truelove, A. Chasin,
   http://www.openp2p.com/pub/a/p2p/2001/07/02/morpheus.html?page=1
6. "The Case for Reslient Overlay Networks", D. Anderson, H. Balakrishnan, F. Kaashoek, R. Morris, Proc. HotOS VIII, May 2001,
   http://nms.lcs.mit.edu/papers/ron-hotos2001.html

7. "A Case For End System Multicast", Y. Chu, S. Rao, H. Zhang, Proceedings of ACM SIGMETRICS , Santa Clara,CA, June 2000, pp 1-12. http://www.cs.cmu.edu/afs/cs/project/cmcl-yhchu/www/Sigmetrics2000/sigmetrics-2000.ps.gz

8. "Enabling Conferencing Applications on the Internet Using an Overlay Multicast Architecture", Y. Chu, S. Rao, S. Seshan, H. Zhang, Proc. ACM Sigcomm 2001, http://www.acm.org/sigs/sigcomm/sigcomm2001/p5-chu.pdf

9. "Overcast: Reliable Multicasting with an Overlay Network", J. Jannotti, D. K. Gifford, K. L. Johnson, M. F. Kaashoek, and J. W. O'Toole, Jr., Proceedings of OSDI'00. http://gaia.cs.umass.edu/cs791n/Jannotti00.pdf

10. "Tapestry: A Fault Tolerant Wide Area Network Infrastructure", B. Zhou, D. A. Joseph, J. Kubiatowicz, Sigcomm 2001 poster and UC Berkeley Tech. Report UCB/CSD-01-1141. http://www.cs.berkeley.edu/~ravenben/publications/CSD-01-1141.pdf

11. "Chord: A Scalable Peer-To-Peer Lookup Service for Internet Applications", I. Stoica, R. Morris, D. Karger, F. Kaashoek, H. Balakrishnan, ACM Sigcomm 2001, http://www.acm.org/sigcomm/sigcomm2001/p12.html

12. "A Scalable Content-Addressable Network", S. Ratnasamy, P. Francis, M. Handley, R. Karp, S. Shenker, ACM Sigcomm 2001, http://www.acm.org/sigcomm/sigcomm2001/p13.html

13. "Application-level Multicast using Content-Addressable Networks", Sylvia Ratnasamy, Mark Handley, Richard Karp, Scott Shenker In Proceedings of NGC 2001 http://www.icir.org/sylvia/can-mcast.ps

14. "Application-Level Anycasting: a Server Selection Architecture and Use in a Replicated Web Service", E. Zegura, M. Ammar, Z. Fei, and S. Bhattacharjee. IEEE/ACM Transactions on Networking, Aug. 2000. ftp://ftp.cs.umd.edu/pub/bobby/publications/anycast-ToN-2000.ps.gz

15. "Evaluation of a Novel Two-Step Server Selection Metric", K. M. Hanna, N. Natarajan, and B.N. Levine, in IEEE ICNP 2001. http://www.cs.umass.edu/~hanna/papers/icnp01.ps

16. "Fault Tolerant Replication Management in Large Scale Distributed Storage Systems", R. Golding, E. Borowski, 1999 Symposuium on Reliable Distributed Systems, http://gaia.cs.umass.edu/cs791n/peters_paper.pdf

17. "Xenoservers: Accounted Execution of Untrusted Code", D. Reed, I. Pratt, P. Menage, S. Early, N. Stratford, Proceedings, 7th Workshop on Hot Topics in Operating Systems HotOS 1999 http://www.cl.cam.ac.uk/Research/SRG/netos/xeno/

18. "OceanStore: An Architecture for Global-Scale Persistent Storage", J. Kubiatowicz, et al., in Proceedings of the Ninth international Conference on Architectural Support for Programming Languages and Operating Systems (ASPLOS 2000), November 2000. http://oceanstore.cs.berkeley.edu/publications/papers/pdf/asplos00.pdf

19. "A Measurement Study of Napster and Gnutella as Examples of Peer-Peer File Sharing Systems", P. Gummagi, S Sariou, S. Gribble, ACM Sigcomm 2001 poster.

20. "Free Riding on Gnutella", E. Adar, and B. Huberman, First Monday, Vol. 5, No. 10, http://www.firstmonday.dk/issues/issue5_10/adar/

21. "Search in Power-Law Networks", L. Adamic, A. Lukose, R. Lukose, and B. Huberman, http://www.hpl.hp.com/shl/papers/plsearch/

22. "SCRIBE: The Design of a Large-scale Event Notification Infrastructure", A. Rowstron, A-M. Kermarrec, P. Druschel and M. Castro, Submitted June 2001. http://www.research.microsoft.com/~antr/PAST/scribe.pdf

23. "Security Aspects of Napster and Gnutella", S. Bellovin, Distinguished lecture at
    Polytechnic U., Real-Audio presentation (1 hour).
    http://media.poly.edu/RealMedia/electrical/eesem10_26.ram. Slides available at
    http://www.research.att.com/~smb/talks/NapsterGnutella.
24. "Publius: A Robust, Tamper-evident, Censorship-resistant, Web Publishing Sys-
    tem", Marc Waldman, Aviel D. Rubin and Lorrie Faith Cranor, Proc. 9th USENIX
    Security Symposium, pp 59-72, August 2000.
    http://publius.cdt.org/publius.pdf
25. "Crowds: Anonymity for Web Transactions", M. Reiter and A. Rubin, ACM Trans-
    actions on Information and System Security, November 1998
26. "Responder Anonymity and Anonymous Peer-to-Peer file sharing", V. Scarlata,
    B.N. Levine, and C. Sheilds, ICNP 2001.
    http://signl.cs.umass.edu/pubs/scarlata.apfs.ps.gz
27. "The Small-World Phenomenon: An Algorithmic Perspective", Jon Kleinberg,
    http://www.cs.cornell.edu/home/kleinber/swn.ps
28. "Streaming Live Media over a Peer-to-Peer Network", H. Deshpande, M. Bawa, H.
    Garcia-Molina, http://dbpubs.stanford.edu:8090/pub/2001-31
29. "Storage Management and Caching in PAST, A Large-scale, Persistent Peer-to-
    peer Storage Utility", A. Rowstron, P. Druschel, SOSP 2001.
    http://www-cse.ucsd.edu/sosp01/papers/rowstron.pdf
30. "Routing in Overlay Multicast Networks", S. Shi and J. Turner, Technical Report,
    TR 01-19, Washington University,
    http://www.arl.wustl.edu/arl/Publications/2000-04/wucs0119.pdf
31. "What Can Peer-to-Peer Do for Databases, and Vice Versa?", S. Gribble, A.
    Halevy, Z. Ives, M. Rodrig, D. Suciu, in the Fourth International Workshop on
    the Web and Databases (WebDB '2001).
    http://www.cs.washington.edu/homes/gribble/papers/p2p.ps.gz
32. "A peer-to-peer VOIP network", http://phonebazooka.com/
33. "Resilient Overlay Networks", D. Anderson, H. Balakrishnan, F. Kaashoek, R.
    Morris, Proc. 18th ACM SOSP, Banff, Canada, October 2001.
    http://nms.lcs.mit.edu/papers/ron-sosp2001.html
34. "Using the Small World Model to Improve Freenet Performance", H. Zhang, A.
    Goel, R. Givindran, Proceedings of IEEE Infocom, June 2002.
35. "Peer-to-Peer Architecture Case Study: Gnutella", M. Ripeanu, in proceedings of
    2001 International conference on P2P computing
    http://people.cs.uchicago.edu/~matei/PAPERS/P2P2001.pdf
36. "Determining Characteristics of the Gnutella Network", B. Gedik,
    http://www.cs.gatech.edu/people/home/bgedik/mini-projects/
    mini-project1/FinalReport.htmq
37. Limewire tech papers, http://www.limewire.com/index.jsp/tech_papers
38. Industry viewpoint: http://www.openp2p.com/ and
    http://conferences.oreilly.com/p2p/
39. " Peer-to-peer, Harnessing the Power of Disruptive Technologies", Edited by Andy
    Oram, pub. O'Reilly, Mar 2001, ISBN 0-596-00110
40. "Pastry: Scalable, Distributed Object Location and Routing for Large-scale Peer-
    to-Peer Systems" A. Rowstron and P. Druschel, IFIP/ACM International Confer-
    ence on Distributed Systems Platforms (Middleware), Heidelberg, Germany, pages
    329-350, November, 2001.

41. "Exploiting Network Proximity in Peer-to-Peer Overlay Networks", M. Castro, P. Druschel, Y. C. Hu, A. Rowstron, http://www.research.microsoft.com/~antr/Pastry/pubs.htm, submitted for publication.
42. "Feasibility of a Serverless Distributed File System Deployed on an Existing Set of Desktop PCs", William J. Bolosky, John R. Douceur, David Ely, and Marvin Theimer http://research.microsoft.com/sn/farsite/ ACM SIGMETRICS 2000.
43. "Grapevine: An Exercise in Distributed Computing", Andrew D. Birrell, Roy Levin, Roger M. Needham and Michael D. Schroeder, Communications ACM, vol. 25, no. 4, pp. 260–274, Apr. 1982.
44. "Experience with Grapevine: The Growth of a Distributed System", Michael D. Schroeder, Andrew D. Birrell and Roger M. Needham, ACM Transactions on Computer Systems, vol. 2, no. 1, pp. 3–23, Feb. 1984.
45. "The Eternity Service", R. J. Anderson. In Proceedings of Pragocrypt 96, 1996,
46. "The Magna Carta" http://www.nara.gov/exhall/charters/magnacarta/magmain.html
47. "Middle Box Taxonomy" Brian Carpenter, and Scott Brim, April 2001. Work in progress.
48. "Looking Over the Fence at Networks: A Neighbor's View of Networking Research" Committee on Research Horizons in Networking, Computer Science and Telecommunications Board, National Research Council, USA http://www.nap.edu/books/0309076137/html/

Mobile Computing Middleware

Cecilia Mascolo, Licia Capra, and Wolfgang Emmerich

Dept. of Computer Science
University College London
Gower Street, London, WC1E 6BT, UK
{C.Mascolo|L.Capra|W.Emmerich}@cs.ucl.ac.uk

Abstract. Recent advances in wireless networking technologies and the
growing success of mobile computing devices, such as laptop computers,
third generation mobile phones, personal digital assistants, watches and
the like, are enabling new classes of applications that present challenging problems to designers. Mobile devices face temporary loss of network
connectivity when they move; they are likely to have scarce resources,
such as low battery power, slow CPU speed and little memory; they are
required to react to frequent and unannounced changes in the environment, such as high variability of network bandwidth, and in the resources
availability. To support designers building mobile applications, research
in the field of middleware systems has proliferated. Middleware aims at
facilitating communication and coordination of distributed components,
concealing complexity raised by mobility from application engineers as
much as possible. In this survey, we examine characteristics of mobile
distributed systems and distinguish them from their fixed counterpart.
We introduce a framework and a categorisation of the various middleware systems designed to support mobility, and we present a detailed
and comparative review of the major results reached in this field. An
analysis of current trends inside the mobile middleware community and
a discussion of further directions of research conclude the survey.

1 Introduction

Wireless devices, such as laptop computers, mobile phones, personal digital assistants, smartcards, watches and the like, are gaining wide popularity. Their
computing capabilities are growing quickly, while they are becoming smaller and
smaller, and more and more part of every day life. These devices can be connected to wireless networks of increasing bandwidth, and software development
kits are available that can be used by third parties to develop applications [70].
The combined use of these technologies on personal devices enables people to
access their personal information as well as public resources *anytime* and *anywhere*.

Applications on these types of devices, however, present challenging problems
to designers. Devices face temporary and unannounced loss of network connectivity when they move, connection sessions are usually short, they need to discover
other hosts in an ad-hoc manner; they are likely to have scarce resources, such as

E. Gregori et al. (Eds.): Networking 2002 Tutorials, LNCS 2497, pp. 20–58, 2002.

low battery power, slow CPUs and little memory; they are required to react to frequent changes in the environment, such as change of location or context conditions, variability of network bandwidth, that will remain by orders of magnitude lower than in fixed networks.

When developing distributed applications, designers do not have to deal explicitly with problems related to distribution, such as heterogeneity, scalability, resource sharing, and the like. *Middleware* developed upon network operating systems provides application designers with a higher level of abstraction, hiding the complexity introduced by distribution. Existing middleware technologies, such as transaction-oriented, message-oriented or object-oriented middleware [21] have been built trying to hide distribution as much as possible, so that the system appears as a single integrated computing facility. In other words, distribution becomes *transparent* [3].

These technologies have been designed and are successfully used for stationary distributed systems built with fixed networks. In the following we analyse the aspect that might not suit mobile settings. Firstly, the interaction primitives, such as distributed transactions, object requests or remote procedure calls, assume a stable and constant connection between components. In mobile systems, in contrast, unreachability is the norm rather than an exception. Secondly, synchronous point-to-point communication supported by object-oriented middleware systems, such as CORBA [53], requires the client asking for a service, and the server delivering that service, to be up and running simultaneously. In a mobile environment, it is often the case that client and server hosts are not connected at the same time, because of voluntary disconnections (e.g., to save battery power) or forced disconnection (e.g., no network coverage). Finally, traditional distributed systems assume a stationary execution environment, characterised by stable and high bandwidth, fixed location for every hosts. Recent developments in object oriented middleware have introduced asynchronous primitives in order to allow a more flexible use. As we will see asynchronous primitives could be a better choice in mobile scenarios.

In mobile systems look-up service components are used to hide service location in order to allow reconfiguration with minimal disruption. In mobile systems, where the location of a device changes continuously, and connectivity fluctuates, service and host discovery becomes even more essential, and information on where the services are might have to reach the application layer. While in stationary systems it is reasonable to completely hide context information (e.g., location) and implementation details from the application, in mobile settings it becomes both more difficult and makes little sense. By providing transparency, the middleware must take decisions on behalf of the application. It might, however, be that in constrained and dynamic settings, such as mobile ones, applications can make more efficient and better quality decisions based on application-specific information.

In order to cope with these limitations, many research efforts have focused on designing new middleware systems capable of supporting the requirements imposed by mobility. As a result of these efforts, a pool of mobile middleware systems has been produced. In this survey, we provide a framework and a classification of the most relevant literature in this area, highlighting goals that have

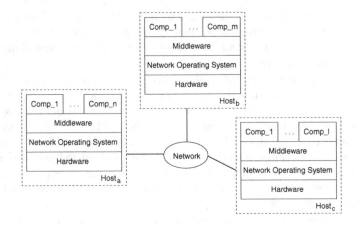

Fig. 1. Example of a distributed system [20].

been attained and goals that still need to be pursued. Our aim is to help middleware practitioners and researchers to categorise, compare and evaluate the relative strengths and limitations of approaches that have been, or might be, applied to this problem. Because exhaustive coverage of all existing and potential approaches is impossible, we attempt to identify key characteristics of existing approaches that cluster them into more or less natural categories. This allows classes of middleware systems, not just instances, to be compared.

Section 2 describes the main characteristics of mobile systems and highlights the many extents to which they differ from fixed distributed systems. Section 3 presents a reference model; Section 4 describes the main characteristics of middleware for distributed systems and their limitations in terms of mobility requirements. Section 5 contains a detailed and comparative review of the major results reached to date, based on the framework presented before. For every class, we give a brief description of the main characteristics, illustrate some examples, and highlight strengths and limitations. Section 6 points out future directions of research in the area of middleware for mobile computing and Section 7 concludes the paper.

2   What Is a Mobile Distributed System?

In this section, we introduce a framework that we will use to highlight the similarities but more importantly the differences between fixed distributed systems and mobile systems. This preliminary discussion is necessary to understand the different requirements that middleware for fixed distributed systems and middleware for mobile systems should satisfy.

2.1 Characterisation of Distributed Systems and Middleware

A distributed system consists of a collection of components, distributed over various computers (also called hosts) connected via a computer network. These components need to interact with each other, in order, for example, to exchange data or to access each other's services. Although this interaction may be built directly on top of network operating system primitives, this would be too complex for many application developers. Instead, a middleware is layered between distributed system components and network operating system components; its task is to facilitate component interactions. Figure 1 illustrates an example of a distributed system.

This definition of distributed system applies to both fixed and mobile systems. To understand the many differences existing between the two, we extrapolate three concepts hidden in the previous definition that greatly influence the type of middleware system adopted: the concept of *device*, of *network connection* and of *execution context*. These concepts are depicted in Figure 2.

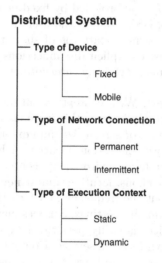

Fig. 2. Characterisation of mobile distributed systems.

Type of Device: as a first basic distinction, devices in a fixed distributed system are *fixed*, while they are *mobile* in a mobile distributed one. This is a key point: fixed devices vary from home PCs, to Unix workstations, to IBM mainframes; mobile devices vary from personal digital assistants, to mobile phones, cameras and smartcards. While the former are generally powerful machines, with large amounts of memory and very fast processors, the latter have limited capabilities, like slow CPU speed, little memory, low battery power and small screen size.

Type of Network Connection: fixed hosts are usually *permanently* connected to the network through continuous high-bandwidth links. Disconnections are either explicitly performed for administrative reasons or are caused by unpredictable failures. These failures are treated as exceptions to the normal behaviour of the system. Such assumptions do not hold for mobile devices that connect to the Internet via wireless links. The performance of wireless networks (i.e., GSM networks and satellite, WaveLAN [30], HiperLAN [47], Bluetooth [10]) may vary depending on the protocols and technologies being used; reasonable bandwidth may be achieved, for instance, if the hosts are within reach of a few (hundreds) meters from their base station, and if they are few in number in the same area, as, for some of the technologies, all different hosts in a cell share the bandwidth, and if they grow, the bandwidth rapidly drops. Moreover, if a device moves to an area with no coverage or with high interference, bandwidth may suddenly drop to zero and the connection may be lost. Unpredictable disconnections cannot be considered as an exception any longer, but they rather become part of normal wireless communication. Some network protocols, such as GSM, have a broader coverage in some areas but provide bandwidth that is smaller by orders of magnitude than the one provided by fixed network protocols (e.g., 9.6 Kbps against 10Gbs). Also, GSM charges the users for the period of time they are connected; this pushes users to patterns of short time connections. Either because of failures or because of explicit disconnections, the network connection of mobile distributed systems is typically *intermittent*.

Type of Execution Context. With context, we mean everything that can influence the behaviour of an application; this includes resources internal to the device, like amount of memory or screen size, and external resources, like bandwidth, quality of the network connection, location or hosts (or services) in the proximity. In a fixed distributed environment, context is more or less *static*: bandwidth is high and continuous, location almost never changes, hosts can be added, deleted or moved, but the frequency at which this happens is by orders of magnitude lower than in mobile settings. Services may change as well, but the discovery of available services is easily performed by forcing service providers to register with a well-known location service. Context is extremely *dynamic* in mobile systems. Hosts may come and leave generally much more rapidly. Service lookup is more complex in the mobile scenario, especially in case the fixed infrastructure is completely missing. Broadcasting is the usual way of implementing service advertisement, however this has to be carefully engineered in order to save the limited resources (e.g., sending and receiving is power consuming), and to avoid flooding the network with messages. Location is no longer fixed: the size of wireless devices has shrunk so much that most of them can be carried in a pocket and moved around easily. Depending on location and mobility, bandwidth and quality of the network connection may vary greatly. For example, if a PDA is equipped with both a WaveLan network card and a GSM module, connection may drop from 10Mbs bandwidth, when close to a base station (e.g., in a conference room) to less than 9.6 Kpbs when we are outdoor in a GSM cell (e.g., in a car on our way home).

According to the type of device, of network connection and of context, typical of distributed systems, we can distinguish: *traditional* distributed systems, *nomadic* distributed systems, and *ad-hoc* mobile distributed systems.

2.2 Traditional Distributed Systems

According to the framework previously described, traditional distributed systems are a collection of fixed hosts, permanently connected to the network via high-bandwidth and stable links, executing in a static environment. Application designers building distributed applications on top of this physical infrastructure (Figure 3) often have to guarantee the following non-functional requirements:

- scalability, that is, the ability to accommodate a higher load at some time in the future. The load can be measured using many different parameters, such as, for instance, the maximum number of concurrent users, the number of transactions executed in a time unit, the data volume that has to be handled;
- openness, that is, the possibility to extend and modify the system easily, as a consequence of changed functional requirements. Any real distributed system will evolve during its lifetime. The system needs to have a stable architecture so that new components can be easily integrated while preserving previous investments;
- heterogeneity, that calls for integration of components written using different programming languages, running on different operating systems, executing on different hardware platforms. In a distributed system, heterogeneity is almost unavoidable, as different components may require different implementation technologies. The ability to establish communication between them is essential;
- fault-tolerance, that is, the ability to recover from faults without halting the whole system. Faults happen because of hardware or software failures (e.g., software errors, ageing hardware, etc), and distributed components must continue to operate even if other components they rely on have failed;
- resource sharing. In a distributed system, hardware and software resources (e.g., a printer, a database, etc.), are shared among the different users of the system; some form of access control of the shared resources is necessary in order to grant access to authorised users of the system only.

Traditional distributed systems were the first form of (fixed) distributed system. Since they started to be investigated and employed about 20 years ago, much research effort has been directed to the solutions of the above mentioned problems. Successful middleware technologies have been designed and implemented. We will briefly review some of them in Section 4.

2.3 Nomadic Distributed Systems

Nomadic systems are, in a sense, a compromise between totally fixed and totally mobile systems. Nomadic systems are usually composed of a set of mobile devices and a core infrastructure with fixed and wired nodes.

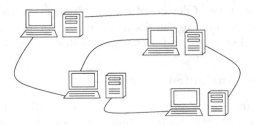

Fig. 3. Structure of a traditional distributed system.

In nomadic systems mobile devices move from location to location, while maintaining a connection to the fixed network. Usually, the wireless network connects the edges of a fixed infrastructure to the mobile devices. The load of computation and connectivity procedures are mainly carried out on the backbone network. Services are mainly provided by the core network to the mobile clients. In some of these, network disconnection is also allowed and services for transparent reconnection and re-synchronisation are provided. The non-functional requirements listed in the section above still hold as the core of these systems is still a fixed network. The scalability of the system is related to the ability of serving larger numbers of mobile devices in an efficient way. Openness has to deal with the extensibility of the functionality provided by the core network. Heterogeneity is complicated by the fact that different links are present (wireless/fixed), and that many different wireless technologies may coexist in the same network. Fault tolerance: depending on the type of application, disconnection may not be a fault or exception but a functionality. Upon reconnection sharing or services should be allowed. Resource sharing, as in fixed networks: most of the resources are on the core network. However, the picture gets more complex if we allow services to be provided by the mobile devices; in this case discovery, quality, and provision need to be thought differently.

2.4   Ad-Hoc Mobile Distributed Systems

Ad-hoc mobile (or simply ad-hoc) distributed systems consist of a set of mobile hosts, connected to the network through high-variable quality links, and executing in an extremely dynamic environment. They differ from traditional and nomadic distributed systems in that they have no fixed infrastructure: mobile hosts can isolate themselves completely and groups may evolve independently, opportunistically forming clusters depicted as clouds in Figure 4 that might eventually rejoin some time in the future. Connectivity may be asymmetric or symmetric depending, for instance, on the radio frequency of the transmission used by the hosts. Radio connectivity defines the clouds depicted in Figure 4 implying that connection is, by default, not transitive. However ad-hoc routing protocols have been defined [51] in order to overcome this limitation and allow routing of packets through mobile hosts.

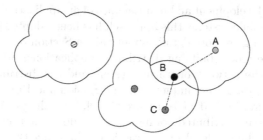

Fig. 4. Structure of an ad-hoc distributed system.

Pure ad-hoc networks have limited applications that range from small ad hoc groups to share information in meetings for a short time to military application on battle-fields and discovery or emergency networks in disastered areas.

The non-functional requirements discussed for nomadic systems still hold in this kind of networks. Scalability becomes an issue when big networks need to be coordinated. In contrast with nomadic systems here every service is provided by a mobile host. In big ad-hoc networks with ad-hoc routing enabled, for instance, routing tables and messages might become big if the network is large. Bandwidth and connectivity may vary depending on concentration and interference among the devices. As for heterogeneity, the network might provide different connectivity strategies and technologies (e.g., Bluetooth areas connected to WaveLan areas) that need coordination. In terms of fault tolerance, given the highly dynamic structure of the network, disconnection has to be consider the norm rather than an exception. Security is even more difficult to obtain than in fixed networks: some message encryption techniques can be used in order to avoid message spoofing.

The physical structure of a mobile ad-hoc network is completely different from the one of traditional fixed networks. In between the two types of network there is a large range of other network solutions which adopt aspects of both. We believe these heterogeneous networks, where fixed components interact with ad-hoc areas, and where different connectivity technologies are used, are going to be the networks of the future and that middleware should provide support for them. Much effort has recently been devolved towards middleware for mobile networks, even if the term mobile network has often been used meaning different things. We will now introduce some of the most significant middleware, together with a categorisation of their peculiarities.

3   Middleware Systems: A Reference Model

Building distributed applications, either mobile or stationary, on top of the network layer is extremely tedious and error-prone. Application developers would have to deal explicitly with all the non-functional requirements listed in the previous section, such as heterogeneity and fault-tolerance, and this complicates considerably the development and maintenance of an application. However, given the novelties of mobile systems this approach has been adopted by many system developers: we will give more details about this in Section 5.

To support designers building distributed applications, middleware system layered between the network operating system and the distributed application is put into place. *Middleware* implements the Session and Presentation Layer of the ISO/OSI Reference Model (see Figure 5) [39]. Its main goal is to enable communication between distributed components. To do so, it provides application developers with a higher level of abstraction built using the primitives of the network operating system. Middleware also offers solutions to resource sharing and fault tolerance requirements.

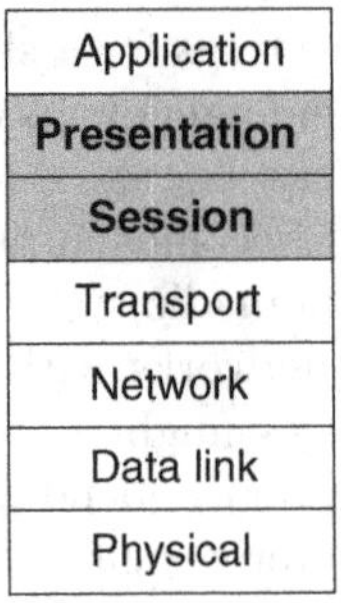

Application
Presentation
Session
Transport
Network
Data link
Physical

Fig. 5. The ISO/OSI Reference Model.

During the past years, middleware technologies for distributed systems have been built and successfully used in industry. For example, object-oriented technologies like OMG CORBA [53], Microsoft COM [58] and Sun Java/RMI [52], or message-oriented technologies like IBM MQSeries [55]. Although very successful in fixed environments, these systems might not to be suitable in a mobile setting, given the different requirements that they entail. We will discuss this issue in more details in Section 4, and show how traditional middleware has been adapted for use in mobile setting in Section 5. Researchers have been and are actively working to design middleware targeted to the mobile setting, and many different solutions have been investigated recently. In order to classify, discuss and compare middleware developed to date, and understand their suitability in a mobile setting, we introduce the reference model depicted in Figure 6.

As shown in the picture, we distinguish middleware systems based on the *computational load* they require to execute, on the *communication paradigms* they support, and on the kind of *context representation* they provide to the applications.

Type of Computational Load. The computational load depends on the set of non-functional requirements met by the system. For instance, the main

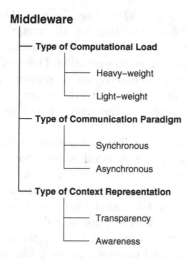

Fig. 6. Characterisation of middleware systems.

purpose of any middleware is to enable communication, e.g., allowing a user to request a remote service; what distinguishes different middleware systems, is the reliability with which these requests are handled. It is in fact much more expensive to guarantee that a request will be executed *exactly once*, instead of providing only *best-effort* reliability, that is, that the request may or may not be executed. As another example, we may consider replication. Replication is widely used by middleware systems in order to achieve fault tolerance and to improve scalability. Keeping the replicas synchronised with the master copy, however, requires a lot of effort and resources (e.g., network communication). Depending on how consistent the replicas are, the computational load of middleware varies accordingly. We use the term *heavy-weight* to denote a system that requires a large amount of resources to deliver services to the above applications, as opposed to a *light-weight* one, which runs using a minimum set of resources (e.g., CPU, main memory, lines of code, etc.). Different computational loads imply different qualities of service and, depending on the characteristics of the distributed system we are addressing (e.g., amount of resources available, non-functional requirements, etc.), different middleware may suit better than others, as we will discuss in Section 3.1 and 3.2.

Type of Communication Paradigm. There are basically two types of communication paradigms a middleware system can support: *synchronous* or *asynchronous*. The former requires that both the client asking for a service, and the server exporting that particular service, are connected and executing at the same time, in order for the request to be successfully processed. The latter, instead, does not require the sender and the receiver of a request to be connected simultaneously. Other forms of communications exist that sit in between these two: *one-way* and *deferred synchronous*. One-way requests return control to the client, without awaiting for the completion of the operation requested from the server. This implies that the semantics of the client does not depend on the result of the requested operation. Deferred synchronous requests return control to the client immediately; unlike one-way requests, however, the client is in charge of re-synchronising with the server to collect the result, and this may cause the client to block if the answer is not yet ready.

Type of Context Representation. The last parameter that we identified to distinguish between different classes of middleware refers to the fact that either information about the execution context is fed to the above applications (i.e., *awareness*) or it is kept hidden inside the middleware itself (i.e., *transparency*). The middleware interacts with the underlying network operating system and collects information about the actual location of a device, value of network bandwidth, latency, remote services available, etc. By transparency, we mean that this context information is used privately by the middleware and not shown to the above applications; for example, the middleware may discover a congestion in a portion of the distributed system and therefore redirect requests to access data to a replica residing on another part of the distributed system, without informing the application about this decision. By awareness, instead, we mean that information about the execution context (or parts of it) is passed up to the running applications, that are now in charge of taking strategic decisions. It would, for example, be the application layer to choose which replica to contact in the non-congestioned portion of the network. Due to the complexity introduced by the awareness approach, middleware for distributed systems usually chose transparent strategies. This is justifiable as the lack in behavioural optimisation with application awareness is compensated by the abundance of resources.

Now that we have discussed a model to characterise distribute systems (Section 2) and a model to characterise middleware for distributed systems (Section 3), we need to understand the relationships between the two models and determine the characteristics that middleware for mobile distributed systems should satisfy and how they differ from the ones required in fixed distributed systems.

3.1   Middleware for Fixed Distributed Systems

With respect to the conceptual model presented above, middleware for fixed distributed systems can be mainly described as resource-consuming systems that hide most of the details of distribution from application designers. With the exception of message-oriented middleware (Section 4.2), they mainly support synchronous communication between components as the basic interaction paradigm. We now analyse in more details the relationship between the physical structure of fixed distributed systems and the characteristics of associated middleware.

Fixed Devices → Heavy Computational Load. As discussed in Section 2, wired distributed systems consist of resource-rich fixed devices. When building distributed applications on top of this infrastructure, it is worthwhile exploiting all the resources available (e.g., fast processors, large amounts of memory, etc.) in order to deliver the best quality of service to the application. The higher the quality of service, the heavier the middleware running underneath the application. This is due to the set of non-functional requirements that the middleware achieves, like fault tolerance, security or resource sharing.

Permanent Connection → Synchronous Communication. Fixed distributed systems are often permanently connected to the network through high-bandwidth and stable links. This means that the sender of a request and its receiver (i.e., the component asking for a service and the component delivering that service) are usually connected at the same time. Permanent connection allows therefore a synchronous form of communication, as the situations when client and server are not connected at the same time are considered only exceptions due to failures of the system (e.g., disconnection due to network overload).

Asynchronous communication mechanisms are however also provided by message oriented middleware (MOM) and recently also by the CORBA specification. Asynchronous communication is used also in fixed networks, however the bulk of middleware applications have been developed using synchronous communication.

Static Context → Transparency. The execution context of a fixed distributed system is generally static: the location of a device seldom changes, the topology of the system is preserved over time, bandwidth remains stable, etc. The abundance of resources allows the disregard of application specific behaviours in favour of a transparent and still efficient approach. For example, to achieve fault tolerance, the middleware can transparently decide on which hosts to create replicas of data and where to redirect requests to access that data in case a network failure inhibits direct access to the master copy, in a completely transparent manner. Hiding context information inside the middleware eases the burden of application programmers that do not have to deal with the achievement of non-functional requirements (e.g., fault tolerance) explicitly, concentrating, instead, on the real problems of the application they are building.

3.2   Middleware for Ad-Hoc and Nomadic Distributed Systems

Nomadic and ad-hoc mobile systems differ in some aspects, however they present a set of similar characteristics that influence the way middleware should behave. We now justify a set of choices which are generally made by the middleware that we will describe.

Mobile Devices → Light Computational Load. Mobile applications run on resource-scarce devices, with little memory, slow CPU, and generally limited battery power. Due to these resource limitations, heavy-weight middleware systems optimised for powerful machines do not seem to suit mobile scenarios. Therefore, the right trade-off between computational load and non-functional requirements achieved by the middleware needs to be established. An example of this might be to relax the assumption of keeping replicas always synchronised, and allow the existence of diverging replicas that will eventually reconcile, in favour of a lighter-weight middleware. We will see examples of this in some of the middleware described in Section 5.

Intermittent Connection → Asynchronous Communication. Mobile devices connect to the network opportunistically for short periods of time, mainly to access some data or to request a service. Even during these periods, the available bandwidth is, by orders of magnitude, lower than in fixed distributed systems, and it may also suddenly drop to zero in areas with no network coverage. It is often the case that the client asking for a service, and the server delivering that service, are not connected at the same time. In order to allow interaction between components that are not executing along the same time line, an asynchronous form of communication is necessary. For example, it might be possible for a client to ask for a service, disconnect from the network, and collect the result of the request at some point later when able to reconnect.

Dynamic Context → Awareness. Unlike fixed distributed systems, mobile systems execute in an extremely dynamic context. Bandwidth may not be stable, services that are available now may not be there a second later, because, for example, while moving the hand-held device loses connection with the service provider. The high variability (together with the constrained resources) influences the way middleware decides and chooses. The optimisation of the application and middleware behaviour using application and context aware techniques becomes then more important, also given the limited resources.

4   Middleware for Fixed Distributed Systems

Middleware technologies for fixed distributed systems can be classified into three main categories[1]: object-oriented middleware, message-oriented middleware and transaction-oriented middleware. We briefly review the main characteristics of these technologies and assess the feasibility of their application in mobile settings.

4.1   Object-Oriented and Component Middleware

Object-oriented middleware supports communication between distributed objects, that is, a client object requests the execution of an operation from a server object that may reside on another host. This class of middleware systems evolved from Remote Procedure Calls [69] (RPCs): the basic form of interaction is still synchronous, that means the client object issuing a request is blocked until the

---

[1] Other categories can be identified, like middleware for scientific computing, but we omit their discussion here because they are only loosely related to the mobile setting.

server object has returned the response. Products in this category include implementations of OMG CORBA [53], like IONA'S Orbix [7] and Borland's VisiBroker [46], the CORBA Component Model (CCM) [48], Microsoft COM [58], Java/RMI [52] and Enterprise JavaBeans [43]. Despite the great success of these technologies in building fixed distributed systems, their applicability to a mobile setting is rather restricted because of the heavy computational load required to run these systems, the mainly synchronous form of object requests supported, and the principle of transparency that has driven their design and that prevents any forms of awareness. The most recent CORBA specification allows for asynchronous communication, however at the time of writing no implementation of it exists. The use of these systems in mobile context has been investigated and is reported in Section 5.

4.2   Message-Oriented Middleware

Message-oriented middleware supports the communication between distributed components via message-passing: client components send a message containing the request for a service execution and its parameters to a server component across the network and the server may respond with a reply message containing the result of the service execution. Message-oriented middleware supports asynchronous communication in a very natural way, achieving de-coupling of client and server, as requested by mobile systems: the client is able to continue processing as soon as the middleware has accepted the message; eventually the server will send a reply message and the client will be able to collect it at a convenient time. However, these middleware systems require resource-rich devices, especially in terms of amount of memory in which to store persistent queues of messages received but not already processed. Sun's Java Message Queue [44] and IBM's MQSeries [55] are examples of very successful message-oriented middleware for traditional distributed systems. We believe there is scope for use of these middleware in mobile settings, and we will discuss how some adaptation of JMS has recently been ported to mobile.

4.3   Transaction-Oriented Middleware

Transaction-oriented middleware systems are mainly used in architectures where components are database applications. They support transactions involving components that run on distributed hosts: a client component clusters several operations within a transaction that the middleware then transports via the network to the server components in a manner that is transparent to both clients and servers. These middleware support both synchronous and asynchronous communication across heterogeneous hosts and achieve high reliability: as long as the participating servers implement the two-phase-commit protocol, the atomicity property of transactions is guaranteed. However, this causes an undue overhead if there is no need to use transactions. Despite their success in fixed systems, the computational load and the transparency that are typical of this kind of middleware (such as IBM CICS [31] and BEA's Tuxedo [28]), make them look not very suitable for mobile settings.

The classification of middleware for fixed distributed systems in the three categories (i.e., object-oriented, message-oriented and transaction-oriented) discussed above is actually not rigid. There is in fact a trend of merging these categories together, as shown by the CORBA Object Transaction Service, a convergence of object-oriented and transaction-oriented middleware, or by the CORBA Event Service and the publish/subscribe communication à la CCM, a union of object-oriented and message-oriented middleware. We will now assess the impact of traditional middleware in mobile settings and also describe a set of middleware developed specifically for mobile.

5   Middleware for Mobile Distributed Systems

There are different examples of use of traditional middleware systems in the context of mobile computing. We will show some examples of adaptation of object-oriented middleware and message oriented middleware to small and mobile devices. The main problem with the object-oriented approach is that it relies on synchronous communication primitives that do not necessarily suit all the possible mobile system architectures. The computational load of these systems is quite high and the principle of transparency they adhere to does not always fit mobile applications.

As we have seen, the requirements for mobile applications are considerably different from the requirements imposed by fixed distributed applications. Some of the developed systems for mobile environments adopted the radical approach of not having a middleware but rather rely on the application to handle all the services and deal with the non-functional requirements, often using a context-aware approach that allows adaptation to changing context [15]. Sun provides J2ME (Java Micro Edition) [70] which is a basic JVM and development package targeting mobile devices. Microsoft recently matched this with .Net Compact Framework [42], which also has support for XML data and web services connectivity.

However this approach is a non-solution, as it completely relies on application designers for the solution of most of the non-functional requirements middleware should provide, starting from scalability.

On the other hand, recently, some middleware specifically targeting the needs of mobile computing have been devised [59]; assumptions such as scarce resources, and fluctuating connectivity have been made in order to reach lightweight solutions. Some of the approaches however target only one of the mobility aspects: for instance, many location-aware systems have been implemented to allow application to use location information to provide services.

We now describe some of the developed solutions to date, starting from the examples of adaptation of object-oriented middleware such as CORBA or Java Messaging Server to mobile platforms, to solutions targeting completely ad-hoc scenarios.

5.1   Traditional Middleware Applied in Mobile Computing

Object-oriented middleware has been adapted to mobile settings, mainly to make mobile devices inter-operated with existing fixed networks (i.e., nomadic setting). The main challenge in this direction is in terms of software size and protocol suitability, as already mentioned. IIOP (i.e., the Internet Inter-ORB Protocol) is the essential part of CORBA that is needed to allow communication among devices. IIOP has been successfully ported to mobile setting and used as a minimal ORB for mobile devices [27]. IIOP defines the minimum protocol necessary to transfer invocations between ORBs. In ALICE [27] hand-helds with Windows CE and GSM adaptors have been used to provide support for client-server architectures in nomadic environments. An adaptation of IIOP specifically for mobile (i.e., LW-IOP, Light-weight Inter-Orb Protocol) has been devised in the DOLMEN project [57]: caching of unsent data combined with an acknowledgement scheme to face wireless medium unreliability. Also actual names of machines are translated dynamically through a name server, which maintains up-to-date information of the hosts location.

In [56] CORBA and IIOP are used together with the WAP (Wireless Access Protocol) stack [24] in order to allow the use of CORBA services on a fixed network through mobile devices connected through WAP and a gateway. IIOP is used to achieve message exchange.

In general, the synchronous connectivity paradigm introduced by traditional middleware assumes a permanent connectivity that cannot be given as granted in most of the mobile computing scenarios. The above mentioned systems are usually targeted to nomadic settings where hand-offs allow mobile devices to roam while being connected. Some minimal support for disconnection is introduced.

There have been more serious attempts in the direction of using traditional middleware using a sort of semi-asynchronous paradigm. Some involved RPC based middleware enhanced with queueing delaying or buffering capabilities in order to cope with intermittent disconnections. Example of these behaviours are Rover [32] or Mobile DCE [65]. As we write, an implementation of the message oriented middleware JMS (Java Messaging Server) has been released [68]. It supports point to point and publish/subscribe models, that is a device can either communicate with a single otehr (through its queue), or register on a topic an be notified of all the messages sent to that topic. We believe this is a good answer to the need for adaptation of traditional middleware to mobile and that the use of publish/subscribe and message oriented systems will be taken further as they offer an asynchronous communication mechanism that allows for disconnected operations. However, communication is not the only aspect that mobile computing middleware should tackle: other important aspects such as context awareness and data sharing need to be addressed.

In the existing examples of use of traditional middleware on mobile, the focus is on provision of services from a back-bone network to a set of mobile devices: the main concerns in this scenarios are connectivity and message exchange. In case of a less structured network or in case services must be provided by mobile devices, traditional middleware paradigms seems to be less suitable and a new set of strategies needs to be used.

The importance of monitoring the condition of the environment, and adaptation to application needs, maybe through communication of context information to the upper layers, becomes vital to achieve reasonable quality of service.

Given the highly dynamic environment and the scarce resources, quality of service provision presents higher challenges in mobile computing. Nevertheless, researchers have devised a number of interesting approaches to quality of service provision to mobile devices [15]. Most of the time the devices are considered terminal nodes and the clients of the service provision, and the network connectivity is assumed fluctuating but almost continuous (like in GSM settings). The probably most significant example of quality of service oriented middleware is Mobiware [2], which uses CORBA, IIOP and Java to allow service quality adaptation in mobile setting. As shown in Figure 7, in Mobiware mobile devices are seen as terminal nodes of the network and the main operations and services are developed on a core programmable network of routers and switches. Mobile devices are connected to access points and can roam from an access point to another.

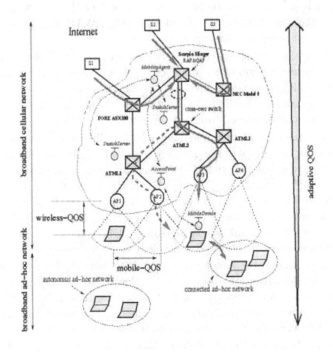

Fig. 7. Mobiware Architecture (from [2]).

The main idea in Mobiware is that mobile devices will have to probe and adapt to the constantly changing resources over the wireless link. The experimental network used by Mobiware is composed of ATM switches, wireless access points, and broadband cellular or ad-hoc connected mobile devices. The toolkit focuses on the delivery of multimedia application to devices with adaptation to the different quality of service and seamless mobility.

Mobiware mostly assumes a service provision scenario where mobile devices are roaming but permanently connected, with fluctuating bandwidth. Even in the case of the ad-hoc broadband link, the device is supposed to receive the service provision from the core network through, first the cellular links and then some ad-hoc hops.

In more extreme scenarios, where links are all ad-hoc, these assumptions cannot be made and different middleware technologies need to be applied. One of the strength of Mobiware is the adaptation component to customise quality of service results. It is more and more clear that middleware for mobile devices should not ignore context and that adaptation is a key point, given the limited resources and changing conditions.

Another interesting example of quality of service oriented middleware is L2imbo [17], a tuple space based quality of service aware system. For a description of L2imbo, we refer the reader to Section 5.4, where we describe the advantages of tuple space based models.

5.2 Context-Awareness Based Middleware

To enable applications to adapt to heterogeneity of hosts and networks as well as variations in the user's environment, systems must provide for mobile applications to be aware of the context in which they are being used. Furthermore, context information can be used to optimise application behaviour counter balancing the scarce resource availability.

User's context includes, but is not limited to:

- location, with varying accuracy depending on the positioning system used;
- relative location, such as proximity to printers and databases;
- device characteristics, such as processing power and input devices;
- physical environment, such as noise level and bandwidth;
- user's activity, such as driving a car or sitting in a lecture theatre.

Context-aware computing is not a new computing paradigm; since it was first proposed a decade ago [64], many researchers have studied and developed systems that collect context information, and adapt to changes.

The principle of *Reflection* has often been used to allow dynamic reconfiguration of middleware and has proven useful to offer context-awareness. The concept of reflection was first introduced by Smith in 1982 [67] as a principle that allows a program to access, reason about and alter its own interpretation. Initially, the use of reflection was restricted to the field of programming language design [33]; some years later, reflection has been applied to the field of operating systems [78] and, more recently, distributed systems [41].

Examples of traditional middleware built around the principle of reflection include, but are not limited to, OpenORB [22], OpenCorba [36], dynamicTAO [34],

Blair et al. work [9]. The role of reflection in distributed systems has to do with the introduction of more openness and flexibility into middleware platforms. In standard middleware, the complexity introduced through distribution is handled by means of abstraction. Implementations details are hidden from both users and application designers and encapsulated inside the middleware itself. Although having proved to be successful in building traditional distributed systems, this approach suffers from severe limitations when applied to the mobile setting. Hiding implementation details means that all the complexity is managed internally by the middleware layer; middleware is in charge of taking decisions on behalf of the application, without letting the application influence this choice. This may lead to computationally heavy middleware systems, characterised by large amounts of code and data they use in order to transparently deal with any kind of problems and find the solution that guarantees the best quality of service to the application. Heavyweight systems cannot however run efficiently on a mobile device as it cannot afford such a computational load. Moreover, in a mobile setting it is neither always possible, nor desirable, to hide all the implementation details from the user. The fundamental problem is that by hiding implementation details the middleware has to take decisions on behalf of the application; the application may, however, have vital information that could lead to more efficient or suitable decisions. Both these limitations can be overcome by reflection. A reflective system may bring modifications to itself by means of inspection and/or adaptation. Through inspection, the internal behaviour of a system is exposed, so that it becomes straightforward to insert additional behaviour to monitor the middleware implementation. Through adaptation, the internal behaviour of a system can be dynamically changed, by modification of existing features or by adding new ones. This means that a middleware core with only a minimal set of functionalities can be installed on a mobile device, and then it is the application which is in charge of monitoring and adapting the behaviour of the middleware according to its own needs.

The possibilities opened by this approach are remarkable: light-weight middleware can be built that support context awareness. Context information can be kept by middleware in its internal data structures and, through reflective mechanisms, applications can acquire information about their execution context and tune the middleware behaviour accordingly. No specific communication paradigm is related to the principle of reflection, so this issue is left unspecified and depends on the specific middleware system built.

Some recent approaches have investigated the use of reflection in the context of mobile systems, and used it to offer dynamic context-awareness and adaptation mechanisms [60]. UIC (Universally Interoperable Core) [74] is a minimal reflective middleware that targets mobile devices. UIC is composed of a pluggable set of components that allow developers to specialise the middleware targeting different devices and environments, thus solving heterogeneity issues. The configuration can also be automatically updated both at compile and run time. Personalities can be defined to have a client-side, server-side or both behaviours. Personalities can also define with which server type to interact (i.e., Corba or

Java RMI) as depicted in Figure 8 : single personalities allow the interaction with only one type while multi personalities allow interaction with more than one type. In the case of multi personalities the middleware dynamically chooses the right interaction paradigm. The size of the core goes, for instance, from 16KB for a client-side CORBA personality running on a Palm OS device to 37KB for a client/server CORBA personality running on a Windows CE device.

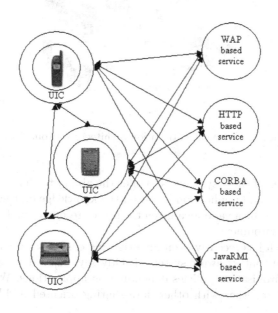

Fig. 8. The UIC Interaction Paradigm (from [60]).

On top of a framework very similar to UIC, Gaia [14] has been developed adding support for dynamic adaptation to context conditions. Gaia defines active spaces where services, users, data and locations are represented and manipulated dynamically and in coordination. Gaia defines a set of further services such as event distribution, discovery, security and data storage. Some other approaches in this direction have been developed focusing on meta-data representation for services and application depended policies. Mechanisms for dynamic adaptation and conflict resolutions have also been put in place [35].

In [76] another approach to context-awareness in mobile computing is presented. The paper presents a middleware for event notification to mobile computing applications. Different event channels allow differentiation of notification of different context and environmental variables. Figure 9 shows the architecture of the system. The asynchronous event channel suits the mobile computing setting allowing for disconnection; however the model the authors assume is based on permanent but fluctuating connectivity. Applications can register for notification of specific events depending on the task they have to perform. Applications relying on this middleware are constructed decoupling the application functionalities from the application context-awareness, where policies are defined to adapt to context.

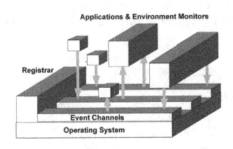

Fig. 9. The event notification architecture (from [76]).

The main idea of these systems is to offer the ability to change the behaviour of the middleware and application based on the knowledge on the changing context. This seems to be a valid idea given the scarce resources and the dynamicity of the mobile environment.

Much research has recently been investigating context-awareness issues. Another example of context-aware systems is Odyssey [62], a data-sharing middleware with synchronisation policies depending on applications. We will describe this middleware together with other data-sharing oriented middleware in Section 5.3.

Location-aware Middleware. Location is one of the most studied aspects of context awareness. Location awareness has attracted a great deal of attention and many examples exist of applications that exploit location information to: offer travellers directional guidance, such as the Shopping Assistant [6] and CyberGuide [37]; to find out neighbouring devices and the services they provide, such as Teleporting [8]; to send advertisements depending on user's location, such as People and Object Pager [12]; to send messages to anyone in a specific area, such as Conference Assistant [19]; and so on. Most of these systems interact directly with the underlying network OS to extract location information, process it, and present it in a convenient format to the user. One of their major limitations concerns the fact that they do not cope with heterogeneity of coordinate information, and therefore different versions have to be released that are able to interact with specific sensor technologies, such as the Global Positioning System (GPS) outdoors, and infrared and radio frequency indoors.

To enhance the development of location-based services and applications, and reduce their development cycle, middleware systems have been built that integrate different positioning technologies by providing a common interface to the different positioning systems. Examples include Oracle iASWE [49], Nexus [25], Alternis [1], SignalSoft [66], CellPoint [13], and many others are being released.

We describe briefly *Nexus* [25], an example of middleware that supports location-aware applications with mobile users. The idea that has motivated the development of this system is that no migration to an homogeneous communication environment is possible, and therefore an infrastructure that supports a heterogeneous communication environment is necessary. Nexus aims to provide this infrastructure.

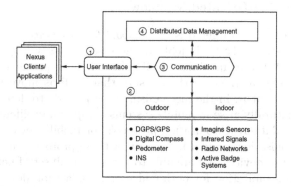

Fig. 10. Nexus Architecture.

The architecture of the Nexus infrastructure is depicted in Figure 10. As the picture shows, there are four different components working together.

The User Interface component is running on the mobile device carried by the user and contains basic functionality, which is required by Nexus Clients to interact with the Nexus platform, and to display and navigate through the model. It also provides support for adapting to devices with different levels of computing power, different amounts of memory, different levels of network connection or different displays.

The interior of a Nexus platform is then split into three main elements: communication, distributed data management and sensors.

Sensors: Nexus applications run both in outdoor and indoor areas. It would be difficult to use only one sensor for positioning in both environments (e.g., GPS can be used outdoor but not indoor, as its satellite signals are blocked by buildings). Therefore, a multi-sensor tool is needed, based on different positioning systems.

Communication: To access information, mobile devices need to be able to connect to the information source, e.g. the Internet, using wireless communication. For a wide area network, data services of mobile telephone systems, such as GSM or UMTS, can be used. Inside a building, wireless LAN, such as Bluetooth, can be used instead. The Nexus communication layer acts as a broker to bridge the differences between existing heterogeneous networks.

Distributed data management: according to the demands of different location aware applications, spatial data have to be offered in multiple representations. Hence, appropriate algorithms to deduce all the necessary levels of detail have to be implemented into the platform. In order to guarantee the interoperability, relationships between different models must be defined and data formats must be exchangeable. All these different aspects concerning the management of data within Nexus are managed by the Distributed data Management component.

5.3   Data Sharing-Oriented Middleware

One of the major issues targeted is the *support for disconnected operations* and data-sharing. Systems like Coda [63], its successor Odyssey [62], Bayou [18,73] and Xmiddle [40] try to maximise availability of data, giving users access to replicas; they differ in the way they ensure that replicas move towards eventual consistency, that is, in the mechanisms they provide to detect and resolve conflicts that naturally arise in mobile systems. Despite a proliferation of different, proprietary data synchronisation protocols for mobile devices, we still lack a single synchronisation standard, as most of these protocols are implemented only on a subset of devices and are able to access a small set of networked data. This represents a limitation for both end users, application developers, service providers and device manufacturers.

Coda
Coda [63] is a file system for large-scale distributed computing environments. It provides resilience to server and network failures through two distinct but complementary mechanisms: server replication and disconnected operation. The first mechanism involves storing copies of a file at multiple servers; the second one is a mode of execution in which a caching site temporarily assumes the role of a replication site; this becomes particularly useful for supporting portable computers.

Coda makes a distinction between relatively few servers, which are physically secure, run trusted software and are monitored by operational staff; and clients, which are far more numerous, may be modified in arbitrary ways by users, are physically dispersed, and may be turned off for long periods of time. The Coda middleware targets therefore fixed or, at most, nomadic systems, as it relies on a fixed core infrastructure.

The unit of replication in Coda is a volume, that is, a collection of files that are stored on one server and form a partial subtree of the shared file system. The set of servers that contain replicas of a volume is its volume storage group (VSG). Disconnections are treated as rare events: when disconnected, the client can access only the data that was previously cached at the client site; upon reconnection modified files and directories from disconnected volumes are propagated to the VSG. Coda clients view disconnected operations as a temporary state and revert to normal operation at the earliest opportunity; these transitions are normally transparent to users. Disconnected operation can also be entered voluntarily, when a client deliberately disconnects from the network; however, clients have no way to influence the portion of the file system that will be replicated locally. Moreover, it is the system that bears the responsibilities of propagating modifications and detecting update conflicts when connectivity is restored. Venus (the cache manager) has control over replication of volumes in use. Venus can be in three states: *Hoarding*, when the client is connected to the network, in this case Venus takes care of caching the information; *Emulating* when disconnected, using the cache and throwing exceptions when the needed data is not in the cache; *Integrating*, upon reconnection, when the data modified need to be reconciled with the server copy. Venus attempts to solve conflicts during reconciliation in an application transparent way; however application specific resolvers (ASR) may be specified.

Odyssey

The mostly application transparent approach adopted by Coda has been improved introducing context-awareness and application-dependent behaviours in Odyssey [62], and allowing the use of these approaches in mobile computing settings. Odyssey, again, assumes that applications reside on mobile clients but access or update data stored on remote, more capable and trustworthy servers; once again the nomadic scenario is targeted.

Odyssey proposes a collaborative model of adaptation. The operating system, as the arbiter of shared resources, is in the best position to determine resource availability; however, the application is the only entity that can properly adapt to given context conditions, and must be allowed to specify adaptation policies. This collaborative model is called application-aware adaptation and it is provided using the architecture depicted in Figure 11.

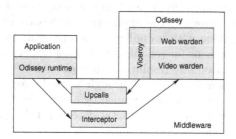

Fig. 11. Odyssey Client Architecture.

To allow reaction to changes, applications first need to register an interest in particular resources. For every resource, they define the acceptable upper and lower bounds on the availability of that resource and they register an up-call procedure that must be invoked whenever the availability of the resource falls outside the window of acceptance. The Viceroy component is then responsible for monitoring resource usage on the client side and notifying applications of significant changes, using the registered up-calls. When an application is notified of a change in resource availability, it must adapt its access. Wardens are the components responsible for implementing the access methods on objects of their type: the interceptor module redirects file systems operations to corresponding wardens, which provide customised behaviour (e.g., different replication policies) according to type-specific knowledge.

Although better suited to the mobile environment than its predecessor Coda, Odyssey suffers from some limitations: the data that can be moved across mobile devices (i.e., a collection of files) may be too coarse-grained in a mobile setting, where devices have limited amount of memory and connection is often expensive and of low quality. Moreover, the existence of a core of more capable and trustworthy servers does not fit the ad-hoc scenario. Finally, files are uninterpreted byte streams; this lack of semantics complicates the development of conflict detection and reconciliation policies from an application point of view.

Bayou

The Bayou storage system [18,73] provides an infrastructure for collaborative applications. Bayou manages conflicts introduced by concurrent activity while relying only on the fluctuating connectivity available in mobile computing. Replication is seen as a requirement in the mobile scenario as a single storage site may not be reachable by some mobile clients or within disconnected work-groups. Bayou allows arbitrary read and write operations to any replica without explicit coordination with the other replicas: every computer eventually receives updates from every other, either directly or indirectly, through a chain of peer interactions. The weak consistency of the replicated data is not transparent to applications; instead, they are aware they may be using weakly consistent data and that their write operations may conflict with those of other users and applications. Moreover, applications are involved in the detection and resolution of conflicts since these naturally depend on the semantics of the application. In particular, the system provides the application with ways of specifying its own notion of a conflict, along with its policy for resolving it. In return, the system implements the mechanisms for reliable detection of conflicts, as specified by the application, and for automatic resolution when possible.

Automatic detection is achieved through a mechanism called *dependency check*: every write operation is accompanied by a dependency set that consists of a query and its expected result. A conflict is detected if the query, when run at a server against its current copy of the data, does not return the expected result. This dependency check is therefore a pre-condition for performing the update. As an example consider someone trying to add an appointment in an agenda, without knowing the content of the agenda. A dependency check could make sure that the time for the appointment is free before adding it. If the check fails, the requested update is not performed and the server invokes a procedure to resolve the detected conflict. Once a conflict has been detected, a *merge procedure* is run by the Bayou server in an attempt to resolve it. Merge procedures are provided by application programmers in the form of templates that are then filled in with the details of each write and accompany each write operation. Users do not have to know about them, except when automatic conflict resolution cannot be done

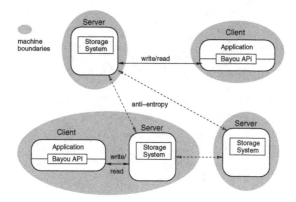

Fig. 12. Bayou System Model.

and manual resolution is needed. In this, Bayou is more flexible than Odyssey as customisation can be performed on each write operation, and not by type.

Bayou guarantees that all the servers will move towards eventual consistency. This means that all the servers will eventually receive all the write operations (through a process called anti-entropy), although the system cannot enforce bounds on write propagation delays since these depend on network connectivity factors that are outside Bayou's control. Eventual consistency is guaranteed by two factors: writes are performed in the same well-defined order on all the servers, and detection and resolution procedures are deterministic so that servers resolve the same conflict in the same way.

Unlike previous systems like Coda, that promote transparency of conflict detection and resolution, Bayou exploits application knowledge for dependency checks and merge procedures. Moreover, while Coda locks complete file volumes when conflicts have been detected but not yet resolved, Bayou allows replicas to always remain accessible. This permits clients to continue to read previously written data and to issue new writes, but it may lead to cascading conflict resolution if the newly issued operations depend on data that are in conflict.

One of the major drawbacks of Bayou is its client-sever architecture. Although in principle client and server may co-exist on a host (see Figure 12), in practise the system requires that each data collection is replicated in full on a number of servers. This is, of course, unaffordable for hand-held devices that can therefore only play the role of clients in this architecture. Bayou is therefore most suited for nomadic rather than ad-hoc applications.

Xmiddle

Xmiddle allows mobile hosts to share data when they are connected, or replicate the data and perform operations on them off-line when they are disconnected; reconciliation of data takes place once the hosts reconnect.

Unlike tuple-space based systems (as we will see in Section 5.4), which store data in flat unstructured structures, Xmiddle allows each device to store its data

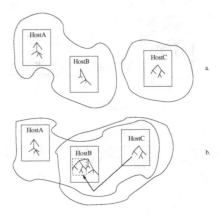

Fig. 13. a. Host H_B and Host H_C are not connected. b. Host H_B and Host H_C connect and Host H_B receives a copy of the tree that it has linked from Host H_C.

in a tree structure. Trees allow sophisticated manipulations due to the different node levels, hierarchy among the nodes, and the relationships among the different elements which could be defined.

When hosts get in touch with each other, they need to be able to interact. Xmiddle allows communication through sharing of trees. On each device, a set of possible access points for the private tree is defined; they essentially address branches of the tree that can be modified and read by peers. The size of these branches can vary from a single node to a complete tree; unlike systems such as Coda and Odyssey (Section 5.3), where entire collections of files have to be replicated, the unit of replication can be easily tuned to accommodate different needs. For example, replication of a full tree can be performed on a laptop, but only of a small branch on a PDA, as the memory capabilities of these devices differ greatly.

In order to share data, a host needs to explicitly *link* to another host's tree. The concept of linking to a tree is similar to the mounting of network file systems in distributed operating systems to access and update information on a remote disk.

Figure 13 shows the general structure of Xmiddle and the way hosts get in touch and interact. As long as two hosts are connected, they can share and modify the information on each other's linked data trees. When disconnections occurs, both explicit (e.g., to save battery power or to perform changes in isolation from other hosts) and implicit (e.g., due to movement of a host into an out of reach area), the disconnected hosts retain replicas of the trees they were sharing while connected, and continue to be able to access and modify the data.

When the two hosts reconnect, the two different, possibly conflicting, replicas need to be reconciled. Xmiddle exploits the tree differencing techniques developed in [72] to detect differences between the replicas which hosts use to concurrently and off-line modify the shared data. However, it may happen that the reconciliation task cannot be completed by the Xmiddle layer alone, because, for example, different updates have been performed on the same node of the tree. In order to solve these conflicts, Xmiddle enables the mobile application engineer to associate application-specific conflict resolution policies to each node of the tree. Whenever a conflict is detected, the reconciliation process finds out which policy the application wants the middleware to apply, in order to successfully complete the merging procedure.

Xmiddle implements the tree data structure using the eXtended Markup Language (XML) [11] and related technologies. In particular, application data are stored in XML documents, which can be semantically associated to trees. Related technologies, such as the Document Object Model (DOM) [4], XPath [16] and XLink [38], are then exploited to manipulate nodes, address branches, and manage references between different parts of an XML document. Reconciliation policies are specified as part of the XML Schema [23] definition of the data structures that are handled by Xmiddle itself.

Xmiddle moves a step forward other middleware systems which focus on the problem of disconnected operations. In particular, unlike Coda and Odyssey, the unit of replication can be easily tuned, in order to accommodate different application and device needs. This issue may play a key role in mobile scenarios, where devices have limited amount of memory and the quality of the network connection is often poor and/or expensive. Moreover, Xmiddle addresses pure ad-hoc networks and not only nomadic ones, as no assumption is made about the existence of more powerful and trusted hosts which should play the role of servers and on which a collection of data should be replicated in full.

Although representing a good starting point for developing middleware for mobile computing, at present Xmiddle suffers from some limitations that require further investigation: the communication paradigm (i.e., sharing of trees) provided is too poor and needs to be improved in order to model more complex interactions that can occur in mobile settings.

5.4   Tuple Space-Based Middleware

The characteristics of wireless communication media (e.g., low and variable bandwidth, frequent disconnections, etc.) favour a decoupled and opportunistic style of communication: decoupled in the sense that computation proceeds even in presence of disconnections, and opportunistic as it exploits connectivity whenever it becomes available. The synchronous communication paradigm supported by many traditional distributed systems has to be replaced by a new asynchronous communication style.

As we have seen, some attempts based on events [76], or queues (Rover [32] or Mobile JMS [68]) have been devised. However, a completely asynchronous and decoupled paradigm (tuple space based) have also been isolated as effective in mobile settings. Although not initially designed for this purpose (their origins go back to Linda [26], a coordination language for concurrent programming), tuple space systems have been shown to provide many useful facilities for communication in wireless settings. In Linda, a tuple space is a globally shared, associatively addressed memory space used by processes to communicate. It acts as a repository (in particular a multi-set) of data structures called tuples that can be seen as vector of typed values. Tuples constitute the basic elements of a tuple space systems; they are created by a process and placed in the tuple space using a `write` primitive, and they can be accessed concurrently by several processes using `read` and `take` primitives, both of which are blocking (even if non-blocking versions can be provided). Tuples are anonymous, thus their selection takes place through pattern matching on the tuple contents. Communications is de-coupled in both time and space: senders and receivers do not need to be available at the same time, because tuples have their own life span, independent of the process that generated them, and mutual knowledge of their location is not necessary for data exchange, as the tuple space looks like a globally shared data space, regardless of machine or platform boundaries.

These forms of decoupling assume enormous importance in a mobile setting, where the parties involved in communication change dynamically due to their migration or connectivity patterns. However, a traditional tuple space implementation is not enough. There are basic questions that need to be answered: how is the globally shared data space presented to mobile hosts? How is it made persistent? The solutions developed to date basically differ depending on the answers they give to the above questions.

We now review three tuple-space middleware that have been devised for mobile computing applications: Lime [45], TSpaces [77] and L2imbo [17].

Lime
In Lime [45], the shift from a fixed context to a dynamically changing one is accomplished by breaking up the Linda tuple space into many tuple spaces, each permanently associated to a mobile unit, and by introducing rules for transient sharing of the individual tuple spaces based on connectivity.

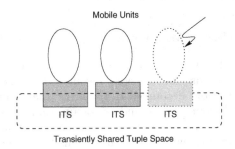

Mobile Units

Transiently Shared Tuple Space

Fig. 14. Transiently shared tuple spaces in Lime.

As shown in Figure 14, each mobile unit has access to an *interface tuple space* (ITS) that is permanently and exclusively attached to that unit and transferred along with it when movement occurs (like in the data tree of Xmiddle). Each ITS contains tuples that the unit wishes to share with others and it represents the only context accessible to the unit when it is alone. Access to the ITS takes place using conventional Linda primitives, whose semantics is basically unaffected. However, the content of the ITS (i.e., the set of tuples that can be accessed through the ITS) is dynamically recomputed in such a way that it looks like the result of the merging of the ITSs of other mobile units currently connected. Upon arrival of a new mobile unit, the content perceived by each mobile unit through its ITS is recomputed taking the content of the new mobile unit into account. This operation is called engagement of tuple spaces; the opposite operation, performed on departure of a mobile unit, is called disengagement. The tuple space that can be accessed through the ITS of a mobile unit is therefore shared by construction and transient because its content changes according to the movement of mobile units.

The term mobile unit can be understood either as mobile agent or as mobile host. In the first case, the context is logical mobility, in the second one, physical mobility. The Lime notion of transiently shared tuple space is applicable to a generic mobile unit, regardless of its nature, as long as a notion of connectivity ruling engagement and disengagement is properly defined.

Lime fosters a style of coordination that reduces the details of mobility and distribution to changes to what is perceived as the local tuple space. This powerful view simplifies application design in many scenarios, relieving the designer from explicitly maintaining a view of the context consistent with changes in the configuration of the system. However, this may be too restrictive in domains where higher degrees of context awareness are needed, for example to control the portion of context that has to be accessed.

Lime tries to cope with this problem, first by extending Linda operations with tuple location parameters that allow to operate on different projections of the transiently shared tuple space. Secondly, information about the system configuration is made available through a read-only transiently shared tuple space called LimeSystem, containing details about the mobile components present in the community and their relationship; finally, reactions can be set on the tuple space, to enable actions to be taken in response to a change in the configuration of the system.

An important aspect of Lime is tuple access and movement; events are used to notify users when a new tuple is available.

TSpaces

TSpaces [77] is an IBM middleware system. The goal of TSpaces is to support communication, computation and data management on hand-held devices. TSpaces is the marriage of tuplespace and database technologies, implemented in Java. The tuplespace component provides a flexible communication model; the database component adds stability, durability, advanced query capabilities and extensive data storage capacity; Java adds instant portability.

The TSpaces design distinguishes between clients and servers. A tuple space exists only on a TSpaces server, while a server may host several spaces. Once a tuple space has been created, a TSpaces client is allowed to perform operations, like read and write, on it. A TSpaces server is a centralised server that listens to client requests: each time a client issues an operation, information is sent to the server that, using a lookup operation, finds out the tuplespace on which the

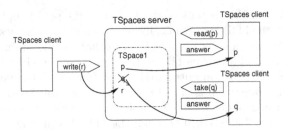

Fig. 15. Examples of interactions in TSpaces.

operation needs to be performed and passes the operation and tuple operand to it to process (see Figure 15).

There are two specific server systems tuple spaces: Galaxy, that contains tuples describing each tuple space that exists on a server; and Admin, that contains access control permissions for each tuplespace and whose goal is to check whether the issuer of each operation has the proper access control privileges.

TSpaces is different from other tuple space based systems for the following reasons: first, the behaviour of the middleware is dynamically modifiable. New operators can be defined, and new datatypes and operators can be introduced. Second, TSpaces employs a real data management layer, with functionalities similar to relational database systems. Operations are performed in a transactional context that ensures the integrity of data. Support for indexing and query capability is provided. Data (i.e., tuples stored in the database) and operations that act on data are kept separated, so that operations can be added or changed without affecting the database.

Unlike Lime, TSpaces mainly targets nomadic environments where servers containing tuple data bases are stored on fixed and powerful machines, reachable by mobile devices roaming around. The transactional approach to tuple read/write is also a limitation in terms of mobility as the paradigms might be too heavy if the connection is fluctuating. Furthermore disconnection is seen as a fault and, when disconnected, clients do not have access to the tuple spaces.

L2imbo. L2imbo [17] is a tuple space based middleware with emphasis on quality of service. Some of the features of L2imbo are: multiple tuple spaces, tuple type hierarchy, quality of service attributes, monitoring and adaptation agents.

Like TSpaces and Lime, L2imbo provides the ability to create multiple tuple spaces. Tuple spaces can be created when needed, provided that creation and termination go through the universal tuple space, which is L2imbo main tuple space. The tuple spaces are implemented in a distributed system fashion, where each host holds a replica of the tuple space, so to allow for disconnected operations. L2imbo also provides quality of service features; quality of service fields can be associated with the tuples, such as deadline for a tuple, which indicates how long a tuple should be available in the space. Associated to these quality of service fields is the QoS monitoring agent, which monitors the conditions of the network, costs of connection and power consumption. Information on quality of service can then be placed into tuples and made available to other agents or hosts.

Given the support for disconnected operations and the use of an asynchronous communication paradigm, L2imbo seems to be well suited for highly mobile environments.

5.5   Service Discovery in Mobile Computing Middleware

In traditional middleware systems, service discovery is provided using fixed name services, which every host knows the existence of. The more the network becomes dynamic, the more difficult service and host discovery becomes. Already in distributed peer-to-peer network [50] service discovery is more complex as hosts join and leave the overlay network very frequently. In mobile systems service discovery can be quite simple, if we speak about nomadic systems where a fixed infrastructure containing all the information and the services is present. However in terms of more ad-hoc or mixed systems where services can be run on roaming hosts, discovery may become very complex and/or expensive.

Most of the ad-hoc systems encountered till now have their own discovery service. Lime and Xmiddle use a completely ad-hoc strategy where hosts continuously monitor their environment to check who is available and what they are offering. A trade-off between power consumption (i.e. broadcast) and discovery needs to be evaluated. Recently, some work on Lime for service advertisement and discovery has been devised [29]. Standard service discovery frameworks have appeared in the recent years: UPnP [75], Jini [5], and Salutation [61]. UPnP stands for Universal Plug and Play and it is an open standard for transparently connecting appliances and services, which is adopted by the Microsoft operating systems. UPnP can work with different protocols such as TCP, SOAP, HTTP. Salutation is a general framework for service discovery, which is platform and OS independent. Jini is instead Java based and dependent on the Java Virtual Machine. The purpose of these frameworks is to allow groups of devices and software components to federate into a single, dynamic distributed system, enabling dynamic discovery of services inside the network federation. We now describe Jini and Salutation.

Jini and JMatos

Jini [5] is a distributed system middleware based on the idea of federating groups of users and resources required by those users. Its main goal is to turn the network into a flexible, easily administered framework on which resources (both hardware devices and software programs) and services can be found, added and deleted by humans and computational clients.

The most important concept within the Jini architecture is the service. A service is an entity that can be used by a person, a program or another service. Members of a Jini system federate in order to share access to services. Services can be found and resolved using a lookup service that maps interfaces indicating the functionality provided by a service to sets of objects that implement that service. The lookup service acts as the central marketplace for offering and finding services by members of the federation. A service is added to a lookup service by a pair of protocols called *discovery* and *join*: the new service provider locates an appropriate lookup service by using the first protocol, and then it joins it, using the second one (see Figure 16). A distributed security model is put in place in order to give access to resources only to authorised users.

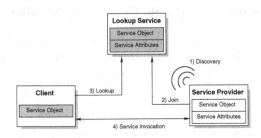

Fig. 16. Discovery, join and lookup mechanism in Jini.

Jini assumes the existence of a fixed infrastructure which provides mechanisms for devices, services and users to join and detach from a network in an easy, natural, often automatic, manner. It relies on the existence of a network of reasonable speed connecting Jini technology-enabled devices.

However the large footprint of Jini (3 Mbytes), due, mainly, to the use of Java RMI, prevents the use of Jini on smaller devices such as iPAQs or PDAs. In this direction Psinaptic JMatos [54] has been developed, complying with the Jini Specification. JMatos does not rely on Java RMI for messaging and has a footprint of just 100 Kbytes.

Salutation
Salutation aims at platform and operating system independence. It focuses on interoperability of different services through a set of common standards for the specification of the service functionalities. The discovery of services is managed by Salutation Managers that interact with each others through RPC, exchanging service registration information. Salutation managers also direct the communication among different clients, notifying them when new data is arriving, for example. The concept of service is decomposed in sets of functional units, representing specific features of the services. Salutation managers also specify the format of the transmitted data implementing transport protocol independence.

In general, despite being the most complete systems in terms of discovery, these systems hardly provide any support of the other important non-functional requirements that would grant the word "middleware" to them. We believe a number of functionalities of the systems we have seen could be combined in an efficient middleware for mobile computing.

6   Future Research Directions

According to our model, a middleware for mobile computing should be lightweight as it must run on hand-held, resource-scarce devices; it should support an asynchronous form of communication, as mobile devices connect to the network opportunistically and for short periods of time; and it should be built with the principle of awareness in mind, to allow its applications to adapt its own and the middleware behaviour to changes in the context of execution, so as to achieve the best quality of service and optimal use of resources.

Furthermore, the variability between different mobile systems, ranging from pure ad-hoc to nomadic with only few roaming hosts, and the different network connectivity paradigms, add further parameters which make it very difficult to isolate a general purpose middleware solution.

We believe research in terms of mobile computing middleware still faces many challenges that might not be solvable adapting traditional middleware techniques.

We believe the future mobile networks will be heterogeneous in the sense that many different devices will be available on the market, with possibly different operating systems and user interfaces. The network connectivity will also be heterogeneous even if an effort towards complete coverage through different connection technologies will be made. For these reasons mobile computing middleware will have to adapt and be customisable in these different dimensions, both at start-up time (i.e., in case of adaptation to different operating systems) and at run-time (i.e., in case of adaptation to different connection technologies).

We also believe application dependent information could play an important role in the adaptation of the behaviour of the middleware and in the trade-off between scarce resource availability and efficient service provision. In this direction, the effort of presentation of the information to the application, and the gathering of application dependent policies is an important presentation layer issue that should be integrated in any mobile computing middleware.

Discovery of existing services is a key point in mobile systems, where the dynamicity of the system is, by orders of magnitude, higher than traditional distributed systems. Recently, interesting research advances in peer to peer systems [71,50] have focused on discovery issues, that might be applicable, at least partially, to mobile settings. However, considerations on the variability of the connection, of the load and of the resources might be different for mobile scenarios. Furthermore, the integration of quality of service consideration into the service advertisement and discovery might enable some optimisation in the service provision.

Another direction of research concerns security. Portable devices are particularly exposed to security attacks as it is so easy to connect to a wireless link. Dynamic customisation techniques seems somehow to worsen the situation. Reflection is a technique for accessing protected internal data structures and it could cause security problems if malicious programs break the protection mechanism and use the reflective capability to disclose, modify or delete data. Security is a major issue for any mobile computing application and therefore proper measures need to be included in the design of any mobile middleware system.

7 Summary

The growing success of mobile computing devices and networking technologies, such as WaveLan [30] and Bluetooth [10], have called for the investigation of new middleware that deal with mobile computing challenges, such as sudden disconnections, scarce resource availability, fast changing environment, etc. During the last years, research has been active in the field of middleware, and a considerable number of new systems has been designed to support this new computational paradigm.

In this survey we have presented a reference model to classify distributed systems into traditional, nomadic and ad-hoc ones. We have highlighted their similarities and mainly their differences. We then listed a set of characteristics that seem to be effective in mobile computing middleware. This list has been helpful to describe some of the unsuitable aspect of traditional middleware systems in a mobile setting, and the principles that have driven towards a new set of middleware for mobile computing

Several classes of middleware for mobile computing (i.e., reflective middleware, tuplespace-based middleware, context-aware middleware, data-oriented) have been identified, illustrated and comparatively discussed. Although addressing some of the issues related to mobility, none of these systems succeed in providing support for all the requirements highlighted by our framework.

Acknowledgements. The authors would like to thank Zuhlke Engineering (UK) Ltd. for supporting Licia Capra; Stefanos Zachariadis and Gian Pietro Picco for comments on an earlier draft of this paper.

References

1. Alternis S.A. Solutions for Location Data Mediation. http://www.alternis.fr/.
2. O. Angin, A. Campbell, M. Kounavis, and R. Liao. The Mobiware Toolkit: Programmable Support for Adaptive Mobile Netwoking. In *Personal Communications Magazine, Special Issue on Adapting to Network and Client Variability.* IEEE Computer Society Press, August 1998.
3. ANSA. The Advanced Network Systems Architecture (ANSA). Reference manual, Architecture Project Management, Castle Hill, Cambridge, UK, 1989.
4. V. Apparao, S. Byrne, M. Champion, S. Isaacs, I. Jacobs, A. L. Hors, G. Nicol, J. Robie, R. Sutor, C. Wilson, and L. Wood. Document Object Model (DOM) Level 1 Specification. W3C Recommendation http://www.w3.org/TR/1998/REC-DOM-Level-1-19981001, World Wide Web Consortium, Oct. 1998.
5. K. Arnold, B. O'Sullivan, R. W. Scheifler, J. Waldo, and A. Wollrath. *The Jini[tm] Specification.* Addison-Wesley, 1999.

6. A. Asthana and M. C. P. Krzyzanowski. An indoor wireless system for personalized shopping assistence. In *Proceedings of IEEE Workshop on Mobile Computing Systems and Applications*, pages 69–74, Santa Cruz, California, Dec. 1994. IEEE Computer Society Press.

7. S. Baker. *Corba Distributed Objects : Using Orbix*. Addison-Wesley, Nov. 1997.

8. F. Bennett, T. Richardson, and A. Harter. Teleporting - making applications mobile. In *Proc. of the IEEE Workshop on Mobile Computing Systems and Applications*, pages 82–84, Santa Cruz, California, Dec. 1994. IEEE Computer Society Press.

9. G. Blair, G. Coulson, P. Robin, and M. Papathomas. An Architecture for Next Generation Middleware. In *Proceedings of Middleware '98*, pages 191–206. Springer Verlag, Sept. 1998.

10. Bluetooth.com. Bluetooth. http://www.bluetooth.com.

11. T. Bray, J. Paoli, and C. M. Sperberg-McQueen. Extensible Markup Language. Recommendation http://www.w3.org/TR/1998/REC-xml-19980210, World Wide Web Consortium, Mar. 1998.

12. P. Brown. Triggering information by context. *Personal Technologies*, 2(1):1–9, Mar. 1998.

13. CellPoint, Inc. The CellPoint System. http://www.cellpt.com/thetechnology2.htm, 2000.

14. R. Cerqueira, C. K. Hess, M. Román, and R. H. Campbell. Gaia: A Development Infrastructure for Active Spaces. In *Workshop on Application Models and Programming Tools for Ubiquitous Computing (held in conjunction with the UBICOMP 2001)*, Sept. 2001.

15. D. Chalmers and M. Sloman. A Survey of Quality of Service in Mobile Computing Environments. *IEEE Communications Surveys*, Second Quarter:2–10, 1999.

16. J. Clark and S. DeRose. XML Path Language (XPath). Technical Report http://www.w3.org/TR/xpath, World Wide Web Consortium, Nov. 1999.

17. N. Davies, A. Friday, S. Wade, and G. Blair. L2imbo: A Distributed Systems Platform for Mobile Computing . *ACM Mobile Networks and Applications (MONET), Special Issue on Protocols and Software Paradigms of Mobile Networks*, 3(2), 1998.

18. A. Demers, K. Petersen, M. Spreitzer, D. Terry, M. Theimer, and B. welch. The Bayou Architecture: Support for Data Sharing among Mobile Users. In *Proceedings of the IEEE Workshop on Mobile Computing Systems and Applications*, pages 2–7, Santa Cruz, California, Dec. 1994.

19. A. Dey, M. Futakawa, D. Salber, and G. Abowd. The Conference Assistant: Combining Context-Awareness with Wearable Computing. In *Proc. of the 3rd International Symposium on Wearable Computers (ISWC '99)*, pages 21–28, San Franfisco, California, Oct. 1999. IEEE Computer Society Press.

20. W. Emmerich. *Engineering Distributed Objects*. John Wiley & Sons, Apr. 2000.

21. W. Emmerich. Software Engineering and Middleware: A Roadmap. In *The Future of Software Engineering - 22nd Int. Conf. on Software Engineering (ICSE2000)*, pages 117–129. ACM Press, May 2000.

22. ExoLab. OpenORB. http://openorb.exolab.org/openorb.html, 2001.

23. D. C. Fallside. XML Schema. Technical Report http://www.w3.org/TR/xmlschema-0/, World Wide Web Consortium, Apr. 2000.

24. W. Forum. Wireless Application Protocol. http://www.fub.it/dolmen/, 2000.

25. D. Fritsch, D. Klinec, and S. Volz. NEXUS - Positioning and Data Management Concepts for Location Aware Applications. In *Proceedings of the 2nd International Symposium on Telegeoprocessing*, pages 171–184, Nice-Sophia-Antipolis, France, 2000.

26. D. Gelernter. Generative Communication in Linda. *ACM Transactions on Programming Languages and Systems*, 7(1):80–112, 1985.
27. M. Haahr, R. Cunningham, and V. Cahill. Supporting CORBA Applications in a Mobile Environment (ALICE). In *5th Int. Conf. on Mobile Computing and Networking (MobiCom)*. ACM Press, August 1999.
28. C. Hall. *Building Client/Server Applications Using TUXEDO*. Wiley, 1996.
29. R. Handorean and G.-C. Roman. Service Provision in Ad Hoc Networks. In *Coordination 2002*. Springer, 2002.
30. G. Held. *Data Over Wireless Networks: Bluetooth, WAP, and Wireless Lans*. McGraw-Hill, Nov. 2000.
31. E. Hudders. *CICS: A Guide to Internal Structure*. Wiley, 1994.
32. A. D. Joseph, J. A. Tauber, and M. F. Kaashoek. Mobile Computing with the Rover Toolkit. *IEEE Transactions on Computers*, 46(3), 1997.
33. G. Kiczales, J. des Rivieres, and D. Borrow. *The Art of the Metaobject Protocol*. The MIT Press, 1991.
34. F. Kon, M. Román, P. Liu, J. Mao, T. Yamane, L. M. aes, and R. Cambpell. Monitoring, Security, and Dynamic Configuration with the *dynamicTAO* Reflective ORB. In *International Conference on Distributed Systems Platforms and Open Distributed Processing (Middleware'2000)*, pages 121–143, New York, Apr. 2000. ACM/IFIP.
35. L. Capra and W. Emmerich and C. Mascolo. A Micro-Economic Approach to Conflict Resolution in Mobile Computing. March 2002. Submitted for Publication.
36. T. Ledoux. OpenCorba: a Reflective Open Broker. In *Reflection'99*, volume 1616 of *LNCS*, pages 197–214, Saint-Malo, France, 1999. Springer.
37. S. Long, R. Kooper, G. Abowd, and C. Atkenson. Rapid prototyping of mobile context-aware applications: the Cyberguide case study. In *Proceedings of the Second Annual International Conference on Mobile Computing and Networking*, pages 97–107, White Plains, NY, Nov. 1996. ACM Press.
38. E. Maler and S. DeRose. XML Linking Language (XLink). Technical Report http://www.w3.org/TR/1998/WD-xlink-19980303, World Wide Web Consortium, Mar. 1998.
39. B. W. Marsden. *Communication Network Protocols: OSI Explained*. Chartwell-Bratt, 1991.
40. C. Mascolo, L. Capra, S. Zachariadis, and W. Emmerich. XMIDDLE: A Data-Sharing Middleware for Mobile Computing. *Int. Journal on Personal and Wireless Communications*, April 2002.
41. J. McAffer. Meta-level architecture support for distributed objects. In *Proceedings of Reflection'96*, pages 39–62, San Francisco, 1996.
42. Microsoft. NET Compact Framework. http://msdn.microsoft.com/vstudio/device/compactfx.asp, 2002.
43. R. Monson-Haefel. *Enterprise Javabeans*. O'Reilly & Associates, Mar. 2000.
44. R. Monson-Haefel, D. A. Chappell, and M. Loukides. *Java Message Service*. O'Reilly & Associates, Dec. 2000.
45. A. L. Murphy, G. P. Picco, and G.-C. Roman. LIME: A Middleware for Physical and Logical Mobility. In *Proceedings of the 21st International Conference on Distributed Computing Systems (ICDCS-21)*, May 2001.
46. V. Natarajan, S. Reich, and B. Vasudevan. *Programming With Visibroker : A Developer's Guide to Visibroker for Java*. John Wiley & Sons, Oct. 2000.
47. E. B. R. A. Networks. ETSI HIPERLAN/2 Standard. http://portal.etsi.org/bran/kta/Hiperlan/hiperlan2.asp.

48. OMG. CORBA Component Model. http://www.omg.org/cgi-bin/doc?orbos/97-06-12, 1997.
49. Oracle Technology Network. Oracle9i Application Server Wireless. http://technet.oracle.com/products/iaswe/content.html, 2000.
50. A. Oram. *Peer-to-Peer: Harnessing the Power of Disruptive Technologies*. O'Reilly, 2001.
51. C. Perkins. *Ad-hoc Networking*. Addison-Wesley, Jan. 2001.
52. E. Pitt and K. McNiff. *Java.rmi : The Remote Method Invocation Guide*. Addison Wesley, June 2001.
53. A. Pope. *The Corba Reference Guide : Understanding the Common Object Request Broker Architecture*. Addison-Wesley, Jan. 1998.
54. Psinaptic. JMatos. http://www.psinaptic.com/, 2001.
55. I. Redbooks. *MQSeries Version 5.1 Administration and Programming Examples*. IBM Corporation, 1999.
56. T. Reinstorf, R. Ruggaber, J. Seitz, and M. Zitterbart. A WAP-based Session Layer Supporting Distributed Applications in Nomadic Environments. In *Int. Conf on Middleware*, pages 56–76. Springer, Nov. 2001.
57. P. Reynolds and R. Brangeon. Service Machine Development for an Open Long-term Mobile and Fixed Network Environment. http://www.fub.it/dolmen/, 1996.
58. D. Rogerson. *Inside COM*. Microsoft Press, 1997.
59. G.-C. Roman, A. L. Murphy, and G. P. Picco. Software Engineering for Mobility: A Roadmap. In *The Future of Software Engineering - 22nd Int. Conf. on Software Engineering (ICSE2000)*, pages 243–258. ACM Press, May 2000.
60. M. Roman, F. Kon, and R. Campbell. Reflective Middleware: From your Desk to your Hand. *IEEE Communications Surveys*, 2(5), 2001.
61. Salutation Consortium. Salutation. http://www.salutation.org/, 1999.
62. M. Satyanarayanan. Mobile Information Access. *IEEE Personal Communications*, 3(1):26–33, Feb. 1996.
63. M. Satyanarayanan, J. Kistler, P. Kumar, M. Okasaki, E. Siegel, and D. Steere. Coda: A Highly Available File System for a Distributed Workstation Environment. *IEEE Transactions on Computers*, 39(4):447–459, Apr. 1990.
64. B. Schilit, N. Adams, and R. Want. Context-Aware Computing Applications. In *Proc. of the Workshop on Mobile Computing Systems and Applications*, pages 85–90, Santa Cruz, CA, Dec. 1994.
65. A. Schill, W. Bellmann, and S. Kummel. System support for mobile distributed applications, 1995.
66. SignalSoft. Wireless Location services. http://www.signalsoftcorp.com/, 2000.
67. B. Smith. Reflection and Semantics in a Procedural Programming Langage. Phd thesis, MIT, Jan. 1982.
68. Softwired. iBus Mobile. http://www.softwired-inc.com/products/mobile/mobile.html, Apr. 2002.
69. W. R. Stevens. *UNIX Network Programming*. Prentice Hall, 1997.
70. Sun Microsystem, Inc. Java Micro Edition. http://java.sun.com/products/j2me/, 2001.
71. Sun Microsystems, Inc. Jxta Initiative. http://www.jxta.org/, 2001.
72. K. Tai. The Tree-to-Tree Correction Problem. *Journal of the ACM*, 29(3):422–433, 1979.
73. D. Terry, M. Theimer, K. Petersen, A. Demers, M. Spreitzer, and C. Hauser. Managing Update Conflicts in Bayou, a Weakly Connected Replicated Storage System. In *Proceedings of the 15th ACM Symposium on Operating Systems Principles (SOSP-15)*, pages 172–183, Cooper Mountain, Colorado, Aug. 1995.

74. Ubi-core. Universally Interoperable Core. http://www.ubi-core.com, 2001.
75. UPnP Forum. Universal Plug and Play. http://www.upnp.org/, 1998.
76. G. Welling and B. Badrinath. An Architecture for Exporting Environment Awareness to Mobile Computing. *IEEE Transactions on Software Engineering*, 24(5):391–400, 1998.
77. P. Wyckoff, S. W. McLaughry, T. J. Lehman, and D. A. Ford. T Spaces. *IBM Systems Journal*, 37(3):454–474, 1998.
78. Y. Yokote. The Apertos reflective operating system: The concept and its implementation. In *Proceedings of OOPSLA '92*, pages 414–434. ACM Press, 1992.

Network Security in the Multicast Framework

Refik Molva and Alain Pannetrat

Institut EURECOM
2229 Route des Crêtes
F-06904 Sophia Antipolis Cedex, FRANCE
{Refik.Molva, Alain.Pannetrat}@eurecom.fr
http:www.eurecom.fr/~nsteam

Abstract. This chapter is an overview of security requirements raised by multicast communications along with a survey of principal solutions. A detailed requirement analysis points out the characteristics of multicast in terms of security and scalability. The differences between unicast and multicast communications with respect to the design of authentication and confidentiality mechanisms are discussed. Main confidentiality proposals are analyzed in a detailed comparison survey based on the security and scalability criteria.

1 Introduction

This chapter addresses network security in the framework of multicast communications. Both *security* and *multicast* encompass a broad number of problems. We have chosen to deal with two most fundamental aspects of security in the context of multicast: *confidentiality* and *authentication* services for multicast applications. We start with an overview of IP-Multicast and we highlight the main security requirements of large scale applications which are build upon multicast mechanisms.

IP-Multicast was introduced by STEVE DEERING [Dee89,Dee91] and describes a set of mechanisms which allow the delivery of a packet to a *set* of recipients rather than just one recipient as opposed to *unicast* communications. The set of recipients that receive the same multicast packets are considered members of a *multicast group*. In the multicast setting, the sender transmits a single copy of a packet and the network takes care of duplicating the packet at proper branching points in order for each recipient to receive a copy of the original packet. With this approach, only one copy of the original packet will be transmitted on most of the underlying network links, which saves resources compared to an equivalent set of unicast communications, as illustrated in the example of figure 1.

2 The Security of Multicast Applications

The integrity and accuracy of multicast routing information is important for the proper functioning of IP-Multicast. Adversaries may inject bogus routing

E. Gregori et al. (Eds.): Networking 2002 Tutorials, LNCS 2497, pp. 59–82, 2002.

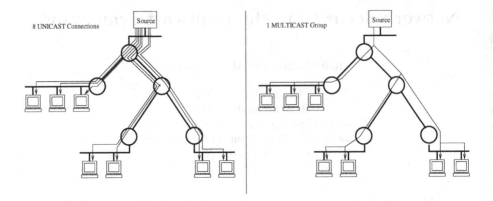

Fig. 1. Unicast or Multicast

messages in the network to modify or even disrupt the routing mechanisms. This type of exposure on routing mechanisms however is not particularly different in multicast. There exist efforts to secure the multicast routing protocols themselves [Moy94,Fen97] and we will not further address this issue in this chapter. The focus of this chapter is geared towards the security of applications which are built on top of IP-Multicast, regardless of lower layer security requirements.

Many large scale commercial applications such as video/audio broadcasting or stock quote distribution could benefit from IP-Multicast mechanisms to reach many receivers in a scalable way. In fact most prominent multicast applications have been confined to the enterprise level or for non-commercial uses such as IETF or NASA broadcastings[MS97]. One of the factors behind the reluctance to the large scale deployment of commercial applications using multicast may well be security. Indeed, securing multicast applications turns out to be a challenge in many situations, mainly because existing unicast security solutions cannot be extended to the multicast setting.

2.1   An Application Scenario

One of the best ways to illustrate the difficulty of securing multicast applications is to use a simple application scenario as an illustration.

Scenario Description. Consider for example a large financial news broadcaster who wants to create a business by providing financial data and market analysis to a large set of customers on the Internet through multicast. These customers are expected to pay for accessing the broadcast content for variable lengths of time, ranging from less than an hour to several days, depending on their needs.

Security Issues. This financial news broadcaster is faced with two main security issues related to the distribution of its commercial content over the Internet:

Confidentiality: the content should only be accessible to the clients who payed
for the service, and only for the duration corresponding to the payment.

Authentication: the content provider needs to prevent its impersonation by another entity who could try to generate content on its behalf, possibly
harming its image and credibility.

Adapting Unicast Solutions. In unicast communications, two efficient techniques are frequently used to address these issues: symmetric
encryption[BDJR97] and MACs (Message Authentications Codes)[BCK96]. To
refer to these mechanisms informally, we will denote the symmetric encryption
of a plaintext message M with a key K as $C \leftarrow \mathcal{E}_K(M)$ and the corresponding
decryption as $M \leftarrow \mathcal{D}_K(C)$. We will write $\sigma \leftarrow \mathcal{T}_{K^{mac}}(M)$ the algorithm which
takes a message M and a MAC key K^{mac} to produce a MAC tag σ and we denote
$\{1, 0\} \leftarrow \mathcal{V}_{K^{mac}}(M, \sigma)$ the corresponding verification algorithm which returns 1
if $\sigma = \mathcal{T}_{K^{mac}}(M)$ and 0 otherwise. Informally speaking, it should be computationally infeasible for an adversary who sees m pairs $\{M_i, \sigma_i\}$ to generate a new
pair $\{M', \sigma'\}$ which verifies $\mathcal{V}_{K^{mac}}(M', \sigma')$ such that $\{M', \sigma'\} \neq \{M_i, \sigma_i\}$ for
$i = 1, ..., m$ without the knowledge of K^{mac}.

Confidentiality. If the multicast data is encrypted with a single key K common
to all recipients who paid for the service then it should prevent others from
accessing the data. However, when the subscription time of one of the clients
R_j expires then K must be updated with a new value K' in order to prevent
the client R_j from further accessing the data. Without loss of generality, we
can assume in our scenario that each client R_i shares a private encryption key
K_i with the source. If K is the only secret shared between the source and the
recipients, in our setting there is no simple way to exclude a client from the
recipient group other than sending: $\mathcal{E}_{K_i}(K')$ for all $i \neq j$. This approach is
clearly unscalable in anything but relatively small groups.

Restricting access to the multicast data through the use of encryption poses
another more subtle problem: containing security exposures. Indeed, in a large
group there is a non negligible chance that one recipient will be compromised and
that its keys will be exposed. Let us suppose for example that the key K we used
above to encrypt data is exposed on a web page or in a public newsgroup; this
allows anyone within the scope of the multicast group to access the data. Unlike
in unicast communications, the adversary does not need to be connected to the
link or to corrupt the routing protocol between the communication endpoints,
since multicast data is forwarded automatically to him if he requests to join the
group. Consequently, we need to provide mechanisms that limit the impact of
such exposures.

Authentication. In the multicast setting, if we construct a Message Authentication Code or MAC[BCK96] for each packet with a common MAC key K^{mac}
shared between the recipient clients and the source, it will disallow non-clients

who do not possess K^{mac} to forge packets that will appear to originate from the source. Since a MAC is computed efficiently using symmetric cryptographic techniques, each packet can be verified independently, thus, lost packets will not affect the ability to authenticate others. But this approach has one big drawback: any client can impersonate as the source, by generating a message M and a valid tag $\sigma = \mathcal{T}_{K^{mac}}(M)$ from the key K^{mac}. Since the MAC key K^{mac} is common to all clients and the source, there is no way to distinguish a packet generated by the source from a packet generated by a client.

An alternative is to have a different key K_i^{mac} for each recipient R_i and append a list of the corresponding MACs to each packet as:

$$M, \sigma_1 = \mathcal{T}_{K_1^{mac}}(M), \sigma_2 = \mathcal{T}_{K_2^{mac}}(M), ..., \sigma_n = \mathcal{T}_{K_n^{mac}}(M)$$

This approach is clearly impractical in a large group since the overhead per packet will increase linearly with the group size.

A third alternative is to replace the MAC with a digital signature (Digital signatures also provide non-repudiation of origin which was not a requirement in our scenario). However, this solution also has its drawbacks because digital signatures are based on asymmetric techniques that introduce a significantly higher computational cost and a non negligible communication overhead, making this alternative impractical in most scenarios.

A Clash between Security and Scalability

Though the previous example is somewhat simplified, it highlights some of the main difficulties associated with the design of security protocols for multicast in large groups. There seems to be an inherent clash between multicast and security. In the case of confidentiality, there is a conflict between multicast scalability mechanisms and security. Indeed, multicast relies on a single group address to identify the set of recipient rather than explicitly listing them. This anonymous identification is the primary mechanism which allows multicast to scale to virtually any group size. On the other hand, confidentiality requires us to identify explicitly the entities which send or receive data in order to provide them with the right keys to access the encrypted multicast data. In the case of authentication, the problem is not related to the group size explicitly but rather to the requirement of an efficient asymmetric mechanism to disallow receivers from impersonating the sender. Additionally, since most multicast protocols are implemented over a best effort channel, these authentication mechanisms need to tolerate losses.

3   Multicast Confidentiality

The sequel of the chapter focuses on multicast confidentiality. We first present a detailed analysis of the problem that helps us draw the main requirements for multicast confidentiality in large groups in terms of security and scalability.

Based on these requirements we then present an overview of existing multicast confidentiality proposals.

To clarify our discussion of encryption in the context of multicast we will use the following naming conventions:

Source: We call *source* an entity that wishes to transmit a certain *content* to a selected set of chosen recipients using multicast.

Content: We always use the words *multicast content* to refer to the actual cleartext data that the source wishes to transmit securely over a multicast channel.

Recipient: We call *recipient* any entity capable of receiving multicast packets from a certain group, regardless of its capacity to actually decrypt the multicast packets to access the content. The only limit to the set of recipients is dictated by multicast routing protocol restrictions, mainly the scope of the multicast group which is limited by the TTL selected by the source.

Member: We call *member*, a selected recipient which has been given cryptographic keys that enable it to access the *content* of the received multicast packets.

Membership Manager: A *membership manager* describes the entity (or entities) which grants or refuses membership to recipients.

The main focus of this section is to analyze and propose methods that allow a dynamic set of members to access the multicast content transmitted by a source while disallowing other recipients to do so.

Typical large scale multicast application scenarios which require confidentiality services are:

- Pay-per-view TV,
- High quality streaming audio,
- The distribution of software updates,
- News feeds and stock quote distribution,

All these scenarios involve one of very few sources and potentially a very large amount of members or clients. In this paper we have chosen to put an emphasis on solutions geared towards these scenarios. In particular we do not deal with dynamic peer groups [BD95,STW96,STW98] which involves secure *n-to-n* communications in much smaller groups through the cooperation of the entities in the group. Though this work deals with a selective broadcasting medium, we do not either aim to provide a solution to the problem of *broadcast encryption* [FN93] where each message is encrypted for an arbitrary and potentially disjoint subset of a larger known group. In particular, in our setting, the set of potential members is not necessarily known a priori. Consequently, we assume at least a minimal unicast back channel between the recipients and a membership manager. This channel is used at least once when a recipient sends a request to a membership manager to become a member of the group.

4 The Security of Dynamic Groups

We qualify the set of members as *dynamic* because we consider the general case where the set of members evolves through time as members are *added* or *removed* from the group during a session. We define *add* and *remove* operations as follows:

Add: When a *recipient* becomes a *member* of the secure multicast group by receiving proper cryptographic access parameters, we say that he is *added* to the group. To be added to a group, recipients needs to contact a membership manager through a secure authenticated unicast channel. If the recipient is allowed to become a member then it receives keys and other related parameters necessary to start accessing the multicast content. The membership policy is application dependent and may be subordinated to other issues such as payment and billing, which are all beyond the scope of this chapter.

Remove: We say that a *member* is removed from the group, when its ability to access the encrypted multicast content is suppressed by the membership manager. A membership manager can decide to remove a member from the group at any time during the session or at least on a certain interval boundary that is application dependent (a packet, a frame, a quantity of bytes or time...).

In many publications, add and remove operations are referred as *join* and *leave* operations, respectively. We prefer the former terminology because it highlights more clearly the fact that providing access to the multicast content is ultimately the decision of a membership manager rather than the recipient itself. We will restrict the terms *join* and *leave* for multicast routing, to describe the fact that an entity requests or relinquishes receiving multicast packets, independently of any security concerns.

All the application scenarios we described in the previous section involve dynamic groups where members are added or removed from the member group. A dynamic set of members implies two security requirements that are not found in traditional secure unicast communications:

backward secrecy and *forward secrecy*.

Backward secrecy defines the requirement that an added member should be able to access multicast content transmitted before it became a member. This means that if a recipient records encrypted multicast data send prior to the time it is added to the group, it should not be able to decrypt it once it becomes a member. Similarly, *forward secrecy*[1] defines the requirement that a member leaving the group should not be able to further access multicast content.

---

[1] The term "forward secrecy" is used here in the context of multicast security and should not be confused with *perfect forward secrecy* in unicast communication, which deals with the security of past sessions when a session key is compromised.

5   Algorithmic Requirements

The forward and backward secrecy requirements have a immediate consequence: the parameters of the encryption algorithm that protects the multicast content must be changed each time a member is added or removed from the group. We must simultaneously allow members to infer the new parameters used to access the multicast data while disallowing other recipients to do so. This parameter change must be done in a scalable way, independently of the group size, which can be a real challenge, as illustrated by the simple financial content provider scenario in section 2.1. The scalability issues related to multicast key management in dynamic groups were first highlighted in the IOLUS[Mit97] framework by S. MITTRA, who identified two generic problems:

1. The "one does not equal n" failure which occurs when the group cannot be treated as a whole but instead as a set of individuals with distinct demands.
2. The "one affects all" failure which occurs when the action of a member affect the whole group.

Our experience with many multicast proposals has prompted us to refine these definitions to propose the following two requirements:

Processing scalability: The cost supported by an individual component, be it the source, an intermediary forwarding component or a member, should scale to any potential group size and membership duration.

Membership robustness: Members should be able to access the content of received multicast packets *as soon* as they are received.

The *processing scalability* requirement was informally illustrated by our sample scenario in section 2.1, where we tried to use unicast techniques in a multicast setting to provide confidentiality. To remove a member from a group of N members it required the source (or the membership manager) to send out a new global key encrypted with $(N-1)$ different private keys to replace the previous global key used to access the content in order to assure forward secrecy. Clearly, this method does not offer processing scalability.

The best effort nature of the Internet is a potential factor that greatly affects membership robustness. Consider a scenario in which the membership manager is required to broadcast a single value R to the whole group each time a member is removed or added, in order for members to update their decryption parameters to continue accessing the multicast content. Several multicast applications can tolerate packet losses. However, if some members experience the loss or the delay of R they will not be able to update the keys needed to access the multicast content, thus the newly received content packets will be worthless. Consequently, in such a scenario, add and remove operations may impair membership robustness.

6   Collusion, Containment, and Trust

While the main goal of multicast encryption in a dynamic group is to come up with a protocol which satisfies both the security requirements and the algorithmic requirements, this is still not quite sufficient to provide security in a *large* group. A factor that is often overlooked in most secure multicast proposals is the potential compromise of a member. Indeed, as the member group becomes larger, the probability of member key exposure increases. In a sufficiently large group, there is no doubt that an exposure will occur. These exposures may be done intentionally by a member: it could send its security parameters to another recipient, or they may be unintentional: a "hacker" may steal the parameters held by a member and publish them in a newsgroup or on a web page. Thus, one of the important concepts we put forward in this chapter is that, since we cannot avoid security exposures, we need to limit the impact of such exposures, a property that we call *containment*.

A second issue is *collusion*: if a few members collude together they should not be able to elevate their privileges beyond the sum of their own privileges. For example, consider a scenario where each member holds secret keys that allow them temporary access to the group. If these members are able to achieve unlimited access by exchanging their secrets and making some computations, then the scheme is clearly weak and will be subverted quickly in a large group.

Some multicast schemes use intermediate elements in the network, such as routers, subgroup servers or proxies as participants in the security protocol. These elements are external to the member group and most likely not always fully controlled by the content provider, the members or a membership manager. In fact, it is quite possible that these intermediary elements will be shared between several secure multicast groups with different member sets. Moreover, in some situations, these elements are numerous enough such that the compromise of one or a few of these entities is probable. Consequently, while it is quite reasonable to assume that the source as well as the few other entities such as a membership manager are secure, this is not realistic for other network resources. Thus we need to *limit the trust* placed in these intermediate elements if they are involved used in security protocols.

7   Requirement Summary

Confidentiality brings out significantly more requirements in the multicast setting than in the unicast setting. We summarize these requirements here as a reference for the analysis of current multicast proposals in the following sections.

Security Requirements:

Requirement 1 Data Confidentiality: *the multicast content should be only accessible to members.*

Requirement 2 Backward and Forward Secrecy: *A new (resp. past) member should not have access to past (resp. future) data exchanged by the group members.*

Requirement 3 Collusion Resistance: *A set of members which exchange their secrets cannot gain additional privileges.*

Requirement 4 Containment: *The compromise of one member should not cause the compromise of the entire group.*

Requirement 5 Limited Intermediary Trust: *Intermediary elements in the network should not be trusted with the security of the group.*

Strictly speaking, the latter requirement (5) is only relevant to protocols which use intermediary elements to actively participate in the security of the group.

Algorithmic Requirements:

Requirement 6 Processing Scalability: *The cost supported by an individual component, be it the source, an intermediary forwarding component or a member, should scale to any potential group size and membership duration.*

Requirement 7 Membership Robustness: *Members should be able to access the content of received multicast packets as soon as they are received.*

8   Overview of Multicast Confidentiality Algorithms

This part of the chapter includes the description of three different multicast confidentiality algorithms: the Logical Key Hierarchy, the re-encryption trees of IOLUS, and the MARKS[Bri99] approach. In the sequel of this chapter, we denote $E_K(x)$ as the encryption of message x with key K with a standard symmetric encryption technique[BDJR97,oST01].

9   LKH: The Logical Key Hierarchy

The Logical Key Hierarchy was independently proposed by WONG ET AL. [WGL98] and WALLNER ET AL.[WHA98].

9.1   Construction

Consider a set of n members $\{R_1, ..., R_n\}$. We build a balanced (or almost balanced) binary tree with n leafs, with a one to one correspondence between a leaf L_i and a member R_i. A random key k_j is then attributed to each vertex j in the tree, including the root and the leafs themselves. This construction that we will call *logical key hierarchy* (LKH) is is illustrated on figure 2 with a small set of 7 recipients.

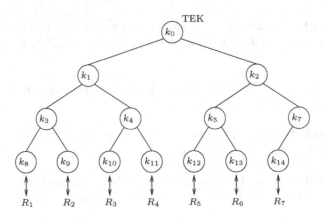

Fig. 2. Basic key graph with 7 members

Setup. Each member R_i receives the set of keys corresponding to the path from the root of the tree to its corresponding leaf L_i. The root of the tree which is known by all members is called the *Traffic Encryption Key* (TEK) and represents the symmetric encryption key used to protect the content distributed over the multicast channel. The remaining keys in the tree are often called *Key Encryption Keys* and are used to update the TEK to guaranty backward and forward secrecy. Referring to our example in figure 2, member R_1 would receive the key set $\{k_0, k_1, k_3, k_8\}$ where k_0 represents the TEK and $\{k_1, k_3, k_8\}$ the KEKs.

9.2   Usage

Let us denote $\mathcal{L}(k_i)$ (respectively $\mathcal{R}(k_i)$) the predicate which returns *true* if the node representing key k_i has a left child (respectively a right child). Further more we will denote $\mathcal{LK}(k_i)$ the key held in the left child of the node representing k_i if it exists, and we similarly define $\mathcal{RK}(k_i)$ for the right child.

To *remove* a member R_i from the group: All keys representing the path from the root to the the leaf L_i corresponding to departing member R_i are invalidated. The leaf L_i corresponding to the departing member is removed from the tree. All the remaining invalidated keys $\{k_1, ..., k_l\}$ in the path from the root to the former leaf are replaced with new random values $\{k'_1, ..., k'_l\}$. For convenience consider $\{k'_1, ..., k'_l\}$ to be ordered by *decreasing* depth in the tree. To allow the remaining members in the tree to update their keys, the membership manager proceeds as follows:

Algorithm 11 *LKH update*
  For $i = 1, ..., (l-1)$ do
    if $\mathcal{L}(k'_i)$ then multicast $E_{\mathcal{LK}(k'_i)}(k'_{(i-1)})$
    if $\mathcal{R}(k'_i)$ then multicast $E_{\mathcal{RK}(k'_i)}(k'_{(i-1)})$

None of the keys that are used for encryption in the previous algorithm are known by the removed member. However, all the remaining members will be able to recover enough keys to update their own key set.

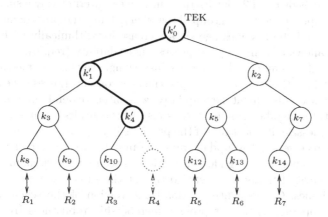

Fig. 3. LKH: A member leaving the group.

As illustrated on figure 3, if member R_4 leaves, then the membership manager needs to broadcast $E_{k_{10}}(k'_4)$, $E_{k_3}(k'_1)$, $E_{k'_4}(k'_1)$, $E_{k'_1}(k'_0)$ and $k_{k_2}(k'_0)$.

To *add* a member to the group: When a member is added to the group, the tree is extended with an additional node. If all leafs in the tree are already attributed we can take a leaf L_i and create two children: the left child is assigned to the member associated with the former leaf L_i and the right child is assigned to the newly added member. Alternatively, we can simply design the tree to be deep enough to accommodate the potential maximum number of members.

Assuming that the new member corresponds to leaf L_i, the sender makes all the keys in the nodes on the path from L_i to the root invalid. A random key is assigned to L_i and transmitted to the added member with a secure unicast channel. All other nodes in the path are updated with the same algorithm that was used above to remove a member from the group.

Key Usage Summary. The scheme distinguishes two type of keys: a TEK and KEK. The TEK is the encryption key that protects the multicast content. It is changed each time a member is added or removed from the group. The KEKs are used to distribute and update the TEK for all members. Until a valid KEK set is explicitly invalidated by the membership manager, it can be maintained in a valid state provided that all update messages are received. Consequently a valid KEK set represents membership in a group and provides long term access to the group, while the TEK itself only represents a short term access to the content.

9.3 Algorithmic Properties of the Scheme

Considering a set of at most n members, the basic LKH scheme requires each recipient to store $\lceil \log_2(n) \rceil + 1$ keys. Each addition or removal of a member requires the broadcast of $2. \lceil \log_2(n) \rceil - 1$ messages (alternatively a single message with all key update messages grouped together). Thus, the processing load of the components and the key message sizes increase logarithmically with the group size, which allows to achieve good processing scalability (req. 6).

However, if a member fails to receive just a single key update message because of a network loss, then it cannot decrypt the new content. Losses have a worsened impact because subsequent key updates will often depend on previous ones. Consequently a member cannot always resynchronizes its parameters on a new key update message if it lost one of the previous key update messages. A member who didn't receive a KEK update message might have to re-register with the membership manager through an authenticated secure unicast channel to receive the proper keys. If many members do this at the same time we may run into scalability issues. Consequently in this basic version of the scheme, both add and remove operations do not provide membership robustness (req. 7) in the presence of losses or strong key update delays.

9.4 Improvements

The basic scheme we described previously has been constantly improved by several authors. We summarize these contributions here.

No-message Member Addition. An interesting improvement was proposed by CARONNI ET AL. [CWSP98] which eliminates the need to multicast a key update message when a member is added to the group. In their proposal, the set $\{k_1, ..., k_l\}$ of invalidated keys represented on the path from the root of the tree are replaced with a set $\{k'_1, ..., k'_l\}$ where each key k'_i is the result of a one way function applied to the previous key, that is $k'_i = F(k_i)$ where F is a publicly known one-way function. This allows members to compute the new keys from the previous ones, and the one-way property of F disallows the new member to compute the previously used keys. The set $\{k'_1, ..., k'_l\}$ is transmitted to the newly added member through a secure unicast channel. In their work CARONNI ET AL. use a counter (a sub-version number) for the TEK and KEKs as ancillary information in the multicast data packets to keep track of the number of times the one-way function was applied (this naturally adds a few bytes of overhead to each packet). Since adding a member does not require the transmission of a multicast message anymore, membership robustness (req. 7) is always assured during this operation, and the problem is now restricted to the removal of a member.

Halving the Update Overhead. MCGREW AND SHERMAN were the first to propose a method to halve the communication overhead of add and remove operations, with a construction called OFT (One Way Function Trees, [MS98]). This further increases the processing scalability of the protocol. We will not describe the OFT construction here, instead we will focus on the more recent ELK approach presented below, which shares some similarity with OFT.

More recently CANNETI ET AL. have proposed a method[CGI+99] that achieves the same reduction and that is compatible with the "no-message-member-addition" we described above. We will describe the core idea behind their protocol here. Assume that f is a one way function, and let $f^k(x)$ denote k successive applications of f to x, where $f^0(x) = x$. As usual, when a member is removed from the group, the remaining keys $\{k_1, ..., k_l\}$ represented on the path from the root to the departing leaf need to be changed to a new set $\{k'_1, ..., k'_l\}$. Here, instead of choosing all these keys at random, the membership manager chooses only one random value r and sets

$$k'_l \leftarrow r, \ k'_{(l-1)} \leftarrow f^1(r), \ k'_{(l-2)} \leftarrow f^2(r), \ ..., \ k'_1 \leftarrow f^{(l-1)}(r)$$

The membership manager then encrypts each k'_i with the key represented by the child node of k'_i that is **not** in the invalidated path. Thus each individual k'_i is encrypted only once as opposed to the original LKH protocol in algorithm 11 which encrypted each key with both child node keys. Because of the relationship between the keys created by f, the nodes can still reconstruct the set of keys they need by applying f.

Example: Recall figure 3 in which member R_4 is removed from the group, the membership manager will compute a random value r and send $E_{k_{10}}(r)$, $E_{k_3}(f(r))$, $E_{k_2}(f(f(r)))$.
  - Member R_3 decrypt $k'_{10} = r$ and computes $k'_4 = f(k'_{10})$, $k'_1 = f(f(k'_{10}))$ as well as $k'_0 = f(f(f(k'_{10})))$.
  - Members R_1 and R_2 decrypt k'_1 and compute $k'_0 = f(k'_1)$.
  - Members R_5, R_6, R_7 decrypt k'_0.

Assuming that the tree is balanced, the communication overhead is at most $\lceil \log_2(n) \rceil$ keys instead of $\lceil 2.\log_2(n) - 1 \rceil$ keys as in the original LKH construction.

Increasing Reliability. Clearly, if the network was perfectly reliable over the whole multicast group the removal of a member would not introduce any membership robustness issues in the LKH scheme (req. 7). However, multicast uses an unreliable transport protocol such as UDP. Consequently, increasing the reliability of the leave operation is crucial if we want to reduce membership robustness issues (req. 7).

WONG AND LAM implemented their LKH[WGL98] protocol over IP-Multicast in Keystone[WL00], and use FEC (Forward Error Correction) to increase the reliability of the addition and removal operations (since they do not use the no-message-member-addition method). Recall that when a member is added or removed from the group in the basic scheme, a set of keys $\{k'_1, k'_2, ..., k'_l\}$

is encrypted with other keys in the tree and transmitted to the group. In the Keystone scheme, these encrypted keys are represented over s packets to which we add r parity packets. The receiver may reconstruct the encrypted key set if he recovers any subset of s packets out of the $(s+r)$ transmitted packets. Since we do not have an absolute guaranty that s packets out of $(s+r)$ packets will always arrive, they combine this mechanism with a unicast re-synchronization mechanism that allows a member to contact a "key cache" server in order to update its keys if losses exceed the correction capacity of the FEC. In their work, they consider a 10-20% loss rate and they show that adding redundancy can substantially increase the likelihood that a member will be able to update its keys. However, as noted by PERRIG ET AL.[PST01] in criticism of the Keystone scheme, the authors assume independent packet losses to compute the key recovery probabilities, while it has been shown that Internet losses occur in bursts.

PERRIG ET AL. proposed a protocol called ELK[PST01] which combines previously proposed ideas such as "no message member addition" and halving the update overhead, with new tricks to increase the reliability of the KEK update operation. The main idea of their work stands in two points:

- First, send a key update message U with reduced communication overhead.
- Second, send many small *hints* that allow a member to recover updated keys if U is lost. The *hints* are smaller than U but require a much higher computational cost from the recipient.

We will give a slightly simplified description of their protocol, omitting some cryptographic key derivation issues that are described in detail in their work[PST01] but are not useful to understand the methods employed here.

Reducing the communication overhead. To reduce the communication overhead of a key update they construct new keys based on contributions from the two children nodes. Consider for example a node containing the key k_i on the path from the root to a leaf representing a removed member which needs to be updated to a new value k_i'. Let k_R (*resp.* k_L) define the key held by the right (*resp. left*) child of the node associated to k_i. Denote $F_k^{\langle p \to m \rangle}(x)$ as a pseudo-random function which takes a p bit value x as input and outputs m bits using a key k. In the ELK construction, the keys in a node are p bits long and are derived from p_1 bits of the left child and p_2 from the right child. To update k_i to k_i' we first derive the right and left contributions C_R and C_L from the old key k_i as follows:

$$C_L \leftarrow F_{k_L}^{\langle p \to p_1 \rangle}(k_i)$$

$$C_R \leftarrow F_{k_R}^{\langle p \to p_2 \rangle}(k_i)$$

Now we compute the new key k_i' by assembling these two contributions to derive k_i':

$$k_i' \leftarrow F^{\langle p \rightarrow p \rangle}_{(C_L || C_R)}(k_i)$$

Consequently, to derive k_i' from k_i the members find themselves in two possible situations:

- They know k_L and thus C_L, and they need to receive C_R to compute k_i', or,
- They know k_R and thus C_R, and they need to receive C_L to compute k_i'.

Consequently, the membership manager will send $E_{k_L}(C_R)$ and $E_{k_R}(C_L)$ to the group where E denotes a stream cipher rather than a block cipher (using the old root key or TEK as Initial Vector). Using a stream cipher allows the membership manager to send $\langle E_{k_L}(C_R), E_{k_R}(C_L) \rangle$ over exactly $p_1 + p_2$ bits. A full key update will thus require the broadcast of $log_2(n)$ chunks of $p_1 + p_2$ bits. Considering $p_1 + p_2 \leq p$ this amounts to having at most 50% of the communication overhead of the original LKH protocol.

Using Hints to increase reliability. Consider the case where $p_1 < p_2$ and where in fact p_1 is small enough such that computing 2^{p_1} symmetric cryptographic computations remains feasible in a reasonably short time. In this situation it's possible to construct an even smaller key update procedure that the authors of ELK call a *hint*, which is defined as the pair $\{V_{k_i'}, E_{k_L}(h)\}$ where $V_{k_i'}$ is a p_3 bit cryptographic checksum computed as follows :

$$V_i' \leftarrow F^{\langle p \rightarrow p_3 \rangle}_{k_i'}(0)$$

With $LSB^{\langle p_2 - p_1 \rangle}(x)$ being the function which returns the $(p_2 - p_1)$ least significant bits of x, h is defined as:

$$h \leftarrow LSB^{\langle p_2 - p_1 \rangle}(F^{\langle n \rightarrow p_2 \rangle}_{k_R}(k_i)) = LSB^{\langle p_2 - p_1 \rangle}(C_R)$$

The hint $\{V_{k_i'}, E_{k_L}(h)\}$ of which the second member h is encrypted with k_L, is broadcasted to the group after the initial key update we described previously. The hint $\{V_{k_i'}, E_{k_L}(h)\}$ can be used to recover the value k_i' in a node. Again, depending on left or right considerations, there are two ways to use the hint:

- Members who know k_L can decrypt h, thus they can recover the $(p_2 - p_1)$ least significant bits of C_R. Since they also know C_L they have to make and exhaustive search for the n_1 missing bits of C_R to recover a value of the form $C_L || \widetilde{C_R}$. For each candidate $C_L || \widetilde{C_R}$ they compute the corresponding $\widetilde{k_i'}$ as $\widetilde{k_i'} \leftarrow F^{\langle p \rightarrow p \rangle}_{(C_L || \widetilde{C_R})}(k_i)$ and use the checksum $V_{k_i'}$ to verify that the candidate is good, otherwise they continue the exhaustive search.
- Members who know k_R will perform an exhaustive search through the 2^{n_1} values that k_L can take. For each potential k_L value they compute the corresponding candidate for k_i'. They use the checksum $V_{k_i'}$ to validate the right candidate.

In practice the authors of ELK suggested in a example to use the values $p_1 = 16$, $p_2 = 35$, $p_3 = 17$ and $n = 64$ using the RC5 cipher[Riv95] both for encryption and also to construct a CBC-MAC based pseudo random function. The authors of ELK implemented this example on a 800Mhz Pentium machine to achieve a high number of computational operations per second.

Aggregating Member Additions and Removals. It should be clear from the description of the LKH algorithm that frequent membership changes increase the workload of all the members of the group and require the membership manager to broadcast a proportional number of key update messages. In situations where the group has a highly dynamic set of member, the membership manager will have to deal with both membership changes and re-synchronization messages for recipients who failed to update their keys correctly, which creates a potential bottleneck. Consequently several authors [CEK+99,LYGL01] have proposed methods to reduce this problem by combining several removal operations into one to achieve a lower communication overhead than the sum of these individual removal operations. In a related but more theoretical approach, [PB99] proposes to organize members in the tree according to their removal probability to increase the efficiency of batch removals and additions.

9.5   LKH: Summary and Conclusion

The LKH construction is an interesting approach, which satisfies the following requirements of multicast confidentiality:

- *Data confidentiality* (req. 1).
- *Backward and forward secrecy* (req. 2).
- *Collusion Resistance*[2] (req. 3).
- *Processing scalability* (req. 6).

Beyond classical multicast forwarding mechanisms, it does not require any additional contribution from the intermediary elements in the network, which clearly also satisfies requirement 5. While some authors have tried to increase the processing scalability of the scheme, the main concern about LKH scheme is membership robustness. Indeed, each time a member is added or removed from the group, all the members need to receive a key update message. If that message is not received by some members, it will disallow them to access the multicast content and it may also disallow them to access further key update messages. The Keystone framework [WL00] uses FEC to increase the likelihood that the key update message will be received by the members, while the ELK framework relies on small "hints" that can be used by a member to derive missing keys in combination with computational tradeoffs.

---

[2] An exception is the *flat key management scheme* presented in [CWSP98], which can be defeated by a coalition of only 2 members.

There remains one requirements that none of the LKH protocol addresses: Containment (req. 4). Indeed, each set of keys held by a member can be used anywhere in the network within the scope of the multicast group to access the encrypted content.

10   Re-encryption Trees

In 1997, MITTRA proposed the Iolus framework, which partitions the group of members into many subsets and creates a tree hierarchy between them. Each subset of members is managed by an distinct entity called a Group Security Intermediary (GSI). The central or top-level subgroup in the hierarchy contains the central group manager called Group Security Center (GSC), as illustrated on the example on figure 4.

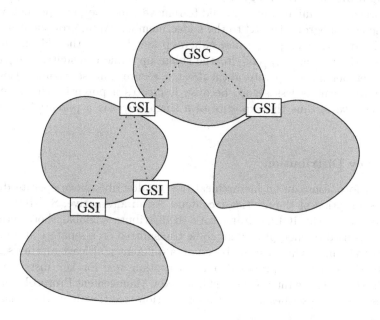

Fig. 4. IOLUS

10.1   Basic Scheme

The GSC produces data encrypted with a key K_1 and multicasts it to a chosen group address G_{A1}. The GSI entities that are direct children of the GSC in the hierarchy then decrypt the data and re-encrypt it with another key K_i, and broadcast it to their own multicast group G_{Ai}. This mechanism is applied recursively all the way down the tree hierarchy. Consequently, this construction can be viewed as a tree network where each node decrypts data received from its parent and re-encrypts it for its children. In this chapter we refer to this type of approach as "re-encryption trees".

Each node or GSI only needs to know the key from its parent in the hierarchy and its own subgroup key. Members in a subgroup are managed by a single GSI and only know the GSI's key K_i.

Removing a member. When a member is removed from a subgroup, the GSI chooses a new subgroup key K_i' and uses several authenticated secure unicast channels to send the new key to the remaining members in the subgroup (alternatively one long multicast message corresponding to the concatenation of these unicast messages).

Adding a member to the group. When a member is added to a subgroup we can proceed in a similar way as we did to remove a member, except that we also send the new subgroup key K_i' to the added member. An improvement in terms of overhead is to encrypt the new subgroup key K_i' with the old one K_i and multicast the message $E_{K_i}(K_i')$ to the subgroup while transmitting K_i' to the added member on a separate authenticated secure unicast channel. However, this improvement is less secure because it creates a potential chain between subsequent keys, thus the exposure of a single key has a potentially stronger impact.

10.2   Key Distribution

The transformations in an hierarchical tree we describe above operate directly on the encryption of the multicast content. As an alternative, S. MITTRA also suggests to use the IOLUS framework to distribute a global short term key. This key is used to encrypt content to be transmitted on a separate autonomous multicast channel, which can be based on a different multicast routing scheme.

HARDJONO ET AL. proposed a re-encryption tree for key distribution in IGKMP[HCD00], the Intra-Domain Group Key Management Protocol. This protocol essentially describes a 2 level re-encryption tree that is used to distribute a common group key $\mathcal{K}$ to the group.

10.3   Analysis

This scheme provides Data Confidentiality (req. 1) as well as backward and forward secrecy (req. 2). Since keys in each subgroup are chosen independently, colluding members do not seem to gain additional privileges beyond the sum of their own access, thus this scheme is collusion resistant (req. 3).

In his work on IOLUS, S. MITTRA did not specify a limit to the size of a subgroup. In practice, however, we need to assume that a subgroup contains no more than M members, otherwise processing scalability issues will arise in subgroups. The benefit of IOLUS is precisely to divide a large multicast group into subgroups that are small enough so that they do not exhibit the scalability issues of a large multicast group. If subgroups are small enough then this scheme will provide processing scalability, since computations and communication overhead will depend on M and not on the full group size.

A second benefit of subgrouping is that it provides stronger membership robustness (req. 7). When a member is added or removed from the group it only affects the other members which depend from the same GSI. Members in other subgroup are not even aware that a member was added or removed from the group. Consequently, the impact of a membership change is restricted to a subgroup of at most M members.

The third benefit of subgrouping is containment (req. 4). Since only at most M members of the same multicast group use the same key K_A to access the multicast data then the exposure of K_A will only have a local impact, limited to the subgroup G_A that uses the same key K_A. If the attacker is in another subgroup G_B, out of scope of G_A than the exposed key K_A will be useless unless he also manages to forward traffic from G_A to G_B.

The main drawback of this scheme is that it fully trusts all the intermediary elements to access the multicast content, which is contrary to requirement 5. Good processing scalability, membership robustness and containment all depend on a reasonably small value of M, the size of the subgroup. But if M is small than it increases the number of Groups Security Intermediaries needed. For large multicast groups, this means that each content provider will need to construct a large network of trusted intermediaries. This adds a significant cost for the content provider.

11 MARKS

The MARKS scheme was proposed by BRISCOE in [Bri99], and features a subscription based approach to multicast confidentiality. Consider a content divided in n segments $\{S_1, ..., S_n\}$ each encrypted with a different key $\{k_1, ..., k_n\}$: the source simply broadcasts $E_{k_i}(S_i)$ for each $1 \leq i \leq n$. To provide access to a member R for segments S_j through $S_{(j+l)}$ the membership manager needs to provide R with the keys $\{k_j, k_{(j+1)}, ..., k_{(j+l)}\}$. Simply sending the list of keys creates an overhead which linearly increases with l, which does not provide processing scalability (req. 6). In his work, BRISCOE provides a method to get around this limitation by constructing a one-way hash tree as follows.

11.1 The Hash Tree

Let n define the number of keys or segments used in the broadcast where $n = 2^d$. Let $\mathcal{L}$ and $\mathcal{R}$ define two one-way functions. For example, we can choose $\mathcal{L}$ (*resp.* $\mathcal{R}$) to be the left (*resp. right*) half of of a pseudo-random generator which doubles its input, or we can use two independently keyed hash functions [BCK96]. We initially choose a random value u_0 and we construct a balanced binary tree as follows:

- assign u_0 to the root.
- if a node is assigned a value u_i we assign to the left child of u_i the value $\mathcal{L}(u_i)$ and to the right child the value $\mathcal{R}(u_i)$.

The leafs of the tree ordered from left to right represent the keys $\{k_1, ..., k_n\}$ used to encrypt the segments $\{S_1, ..., S_n\}$.

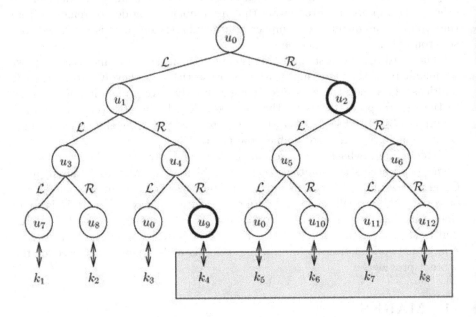

Fig. 5. MARKS with an eight segment stream.

The advantage of this construction is that any sequence $\{k_j, k_{(j+1)}..., k_{(j+l)}\}$ of l consecutive keys in $\{k_1, ..., k_n\}$ can be represented with at most $2.log_2(n) - 1$ values. Indeed, instead of describing a sequence of l keys by listing them in a sequential order, we can simply provide the points in the tree that can be used to derive the needed keys (and only those keys) by applying $\mathcal{R}$ and $\mathcal{L}$. This problem is quite similar to finding IP network aggregations in CIDR[FLYV92]. Consider the example on Figure 5 where a member subscribes for segments 4 to 8. He receives u_2 and u_{10}, next he can efficiently compute $k_4 = u_{10}$, $k_5 = u_{11} = \mathcal{L}(\mathcal{L}(u_2))$, $k_6 = \mathcal{R}(\mathcal{L}(u_2))$, $k_7 = \mathcal{L}(\mathcal{R}(u_2))$, $k_8 = \mathcal{R}(\mathcal{R}(u_2))$.

11.2 Analysis and Conclusion

In the MARKS approach, once the membership manager provides a member with access to l consecutive segments $\mathcal{S} = [S_{j+1}, ..., S_{j+l}]$, it cannot revoke this access unless it redistributes new keys to all other members whose segments intersect with $\mathcal{S}$. In practice it would require the construction of a new key tree over the segments of the stream. Revocation is not possible here without running into strong processing scalability issues. However, there are some multicast applications that will be satisfied with a non revocable subscription based approach. Consider, for example, some Pay-per-view TV applications where the content length is know in advance and the clients pay before the show is broadcasted. For such applications MARKS offers benefits such as:

– Data confidentiality, backward and forward secrecy, collusion resistance and no intermediary trust.
– Processing Scalability.
– Perfect membership robustness.

The last characteristic is the strongest point of MARKS: it does not require any message to be broadcasted to anybody when a recipient's membership expires in the group. When a member is added to the group, he only needs to receives the keys corresponding to his subscription, which can be transmitted to him in advance.

Similarly to LKH, this scheme does not provide any containment. In terms of key exposure, the LKH has a little advantage over MARKS, because the membership can invalidate an exposed key easily, while MARKS by definition does not provide any practical way to do this.

12   Multicast Confidentiality Summary

The three types of multicast confidentiality schemes reviewed in this chapter all suffer from some limitations. The LKH scheme does not offer containment and has some membership robustness issues though some progress has recently been made in that area. The re-encryption tree approach relies on a potentially large infrastructure of fully trusted intermediary elements. Finally the MARKS scheme does not allow member revocation prior to the end of their subscription, neither does it offer any form of containment.

13   Conclusion

Security in multicast raises genuinely new authentication and confidentiality requirements that are not met by classical network security mechanisms. At the core of multicast confidentiality, membership management is a complex issue involving membership managers which communicate on a bidirectional channel with members in different situations. Specific membership management issues are rarely mentioned in multicast confidentiality frameworks, however we believe that these issues would deserve to be explored in future research. For example, in large commercial applications, distributed membership management will become a key requirement. One reason is scalability. Indeed, in a very large multicast group, a single membership manager serving all requests, would quickly become a bottleneck. The second reason is related to laws and regulations. Different countries have different commercial laws and sometimes different cryptography regulations. It makes sense to distribute several membership managers according to these constrains. Re-encryption trees are well suited for distributed membership management, and the MARKS subscription approach is even better. However, it seems that a construction such as LKH is inherently hard to distribute. Besides, multicast confidentiality as described in this chapter is faced with a dilemma. Approaches based on a shared key a common key among members do not offer any form of containment and piracy will greatly hinder the use of such schemes in large groups. On the other hand, the schemes that provide containment require the existence of active intermediate elements in the network with a significant a deployment cost. It is thus not clear yet what type of solution will emerge for confidentiality in large scale multicast applications. Some solutions will possibly rely on a combination of re-encryption trees for inter-domain communications in with an LKH approach for intra-domain communications. Moreover, content providers recently have shown a strong concern in protecting the content they distribute beyond the notion confidentiality we described in this chapter. As in the example of the highly publicized lawsuit[USDC99] that opposed the RIAA to Napster, the mp3 audio download service, issues such as copy protection and copyright have become paramount for content distributors. Architectures that provide these additional services typically rely on several cooperating tamper-proof hardware components achieving end to end protection of the content between the source and the ultimate video or audio playback hardware (see for example, CPRM, `http://www.4Centity.com/tech/cprm/`). Future commercial multicast confidentiality frameworks will potentially rely on such tamper-proof hardware platforms.

References

[BCK96]   M. Bellare, R. Canetti, and H. Krawczyk. Keying hash functions for message authentication. *Lecture Notes in CS*, 1109:1–15, 1996.

[BD95]   M. V. D. Burmester and Y. Desmedt. A secure and efficient conference key distribution system. In Alfredo De Santis, editor, *Advances in Cryptology - EuroCrypt '94*, pages 275–286, Berlin, 1995. Springer-Verlag. Lecture Notes in Computer Science Volume 950.

[BDJR97]   Mihir Bellare, Anand Desai, E. Jokipii, and Phillip Rogaway. A concrete security treatment of symmetric encryption. In *IEEE Symposium on Foundations of Computer Science*, pages 394–403, 1997.

[Bri99]   Bob Briscoe. MARKS: Zero side-effect multicast key management using arbitrarily revealed key sequences. In *First International Workshop on Networked Group Communication*, November 1999.

[CEK+99]   Isabella Chang, Robert Engel, Dilip Kandlur, Dimitrios Pendarakis, and Debanjan Saha. Key management for secure internet multicast using boolean function minimization techniques. In *Proceedings IEEE Infocomm'99*, volume 2, pages 689–698, March 1999.

[CGI+99]   R. Canetti, J. Garay, G. Itkis, D. Micciancio, M. Naor, and B. Pinkas. Multicast security: A taxonomy and some efficient constructions. In *Proceedings of IEEE Infocom'99*, 1999.

[CWSP98]   Germano Caronni, Marcel Waldvogel, Dan Sun, and Berhardt Plattner. Efficient security for large and dynamic multicast groups. In *IEEE 7th Workshop on Enabling Technologies: Infrastructure for Collaborative Enterprises (WET ICE '98)*, 1998.

[Dee89]   Steve E. Deering. RFC 1112: Host extensions for IP multicasting, Aug 1989.

[Dee91]   Steve E. Deering. *Multicast Routing in a Datagram Internetwork*. PhD thesis, Stanford University, 1991.

[Fen97]   W. Fenner. Internet group management protocol, version 2. Request For Comments 2236, November 1997. see also draft-ietf-idmr-igmpv3-and-routing-01.txt for IGMP v3.

[FLYV92]   V. Fuller, T. Li, J. Yu, and K. Varadhan. Supernetting: an address assignment and aggregation strategy, 1992.

[FN93]   A. Fiat and M. Naor. Broadcast encryption. In Douglas R. Stinson, editor, *Advances in Cryptology - Crypto '93*, pages 480–491, Berlin, 1993. Springer-Verlag. Lecture Notes in Computer Science Volume 773.

[HCD00]   Thomas Hardjono, Brad Cain, and Naganand Doraswamy. A framework for group key management for multicast security. Internet-Draft, work in progress, February 2000.

[LYGL01]   Xiaozhou Steve Li, Yang Richard Yang, Mohamed G. Gouda, and Simon S. Lam. Batch rekeying for secure group communications. In *Tenth International World Wide Web conference*, pages 525–534, 2001.

[Mit97]   Suivo Mittra. Iolus: A framework for scalable secure multicasting. In *Proceedings of the ACM SIGCOMM'97 (September 14-18, 1997, Cannes, France)*, 1997.

[Moy94]   John Moy. Multicast extensions to OSPF. Request For Comment 1584, March 1994.

[MS97]   T. Maufer and C. Semeria. Introduction to IP multicast routing. Internet-Draft, July 1997. draft-ietf-mboned-intro-multicast-03.txt.

[MS98]   David A. McGrew and Alan T. Sherman. Key establishment in large dynamic groups using one-way function trees. Technical report, TIS Labs at Network Associates, Inc., Glenwood, MD, 1998.

[oST01]   National Institute of Standards and Technology. Advanced Encryption Standard, 2001.

[PB99]   R. Poovendran and John S. Baras. An information theoretic analysis of rooted-tree based secure multicast key distribution schemes. In *CRYPTO*, pages 624–638, 1999.

[PST01]   Adrian Perrig, Dawn Song, and Doug Tygar. ELK, a new protocol for efficient large-group key distribution. In *IEEE Symposium on Security and Privacy*, May 2001.

[Riv95]   R.L. Rivest. The RC5 encryption algorithm. In In B. Preneel, editor, *Fast Software Encryption*, volume 1008 of *Lecture Notes in Computer Science*, pages 86–96. Springer Verlag, 1995.

[STW96]   Michael Steiner, Gene Tsudik, and Michael Waidner. Diffie-Hellman key distribution extended to group communication. In *Proceedings of the 3rd ACM Conference on Communications Security (March 14-16, 1996, New Delhi, India)*, 1996.

[STW98]   Michael Steiner, Gene Tsudik, and Michael Waidner. CLIQUES: A new approach to group key agreement. In *Proceedings of the 18th International Conference on Distributed Computing Systems (ICDCS'98)*, pages 380–387, Amsterdam, May 1998. IEEECSP.

[USDC99]   Northern District of California United States District Court. RIAA v. Napster, court's 512(a) ruling, 1999.

[WGL98]   C. K. Wong, M. Gouda, and S. S. Lam. Secure group communications using key graphs. In *ACM SIGCOMM 1998*, pages 68–79, 1998.

[WHA98]   Debby M. Wallner, Eric J. Harder, and Ryan C. Agee. Key management for multicast: Issues and architectures. Internet draft, Network working group, september 1998, 1998.

[WL00]   C. Wong and S. Lam. Keystone: a group key management system. In *Proceedings of International Conference in Telecommunications*, 2000.

Categorizing Computing Assets According to Communication Patterns

Dieter Gantenbein[1] and Luca Deri[2]

[1]IBM Zurich Research Laboratory, 8803 Rueschlikon, Switzerland
dga@zurich.ibm.com, http://www.zurich.ibm.com/~dga/

[2]NETikos S.p.A., Via Matteucci 34/b, 56124, Pisa, Italy
deri@ntop.org, http://luca.ntop.org/

Abstract. In today's dynamic information society, organizations critically depend on the underlying computing infrastructure. Tracking computing devices as assets and their usage helps in the provision and maintenance of an efficient, optimized service. A precise understanding of the operational infrastructure and its users also plays a key role during the negotiation of outsourcing contracts and for planning mergers and acquisitions. Building an accurate inventory of computing assets is especially difficult in unknown heterogeneous systems and networking environments without prior device instrumentation. User mobility and mobile, not-always-signed-on, computing devices add to the challenge. We propose to complement basic network-based discovery techniques with the combined log information from network and application servers to compute an aggregate picture of assets, and to categorize their usage with data-mining techniques according to detected communication patterns.

Keywords: Network and System Management, Inventory and Asset Management, IT Audit, Due Diligence, Data Mining, and OLAP.

1 Computing Infrastructure as a Critical Business Resource

Modern e-business environments[1] tightly link the customer and supplier systems with the internal computing infrastructure. Hence the performance of the end-to-end business processes becomes critically dependent on the availability of the underlying computing infrastructure. From an economic perspective, the efficient cost-effective and resource-optimized provision of the required services is an argument in many organizations to justify the tight grip on the deployed computing assets and their usage [16].

Classical methods for asset and inventory management quickly reach their limit in today's dynamic environments: Periodic physical inventories ("wall-to-wall") have the clear advantage of identifying the actual location of the devices but require costly human visits ("sneaker net") and can detect neither mobile, currently out-of-office equipment nor the existence and use of contained logical assets. Financial asset

---

[1] An e-Business definition can be found at
    http://searchebusiness.techtarget.com/sDefinition/0,,sid19_gci212026,00.html

E. Gregori et al. (Eds.): Networking 2002 Tutorials, LNCS 2497, pp. 83-100, 2002.

tracking, while being an accepted process in its own right, cannot detect additional equipment brought into or remotely accessing the resources of an organization. Periodic self-assessment questionnaires to be filled out by individual end users or their cost-center managers are another and often complementary approach. Apart from the human effort they require and the inaccurate incomplete data that results, most forms pose questions the answer of which could be easily read out of the infrastructure itself.

Well-managed computing infrastructures typically equip servers and end-user devices with software daemons for the tracking of resources and the system as well as for application performance monitoring [29][30]. There are many situations, however, in which this cannot be assumed and used. In many organizations, there are a fair number of devices that are brought in ad-hoc and are not instrumented accordingly, for which instrumentation is not available, or on which instrumentation has been disabled. After a merger/acquisition, for example, we can hardly assume to find an encompassing management environment in place across the entire holding organization. However, a good understanding of the provided infrastructure and its users is essential, actually already prior to the acquisition or while negotiating an outsourcing contract. Such investigations to gather the data required allowing more accurate service cost predictions and reducing the risk of unforeseen contractual responsibilities are often called Due Diligence or Joint Verification.

In this paper, we argue that it is no longer sufficient to keep a static inventory of information technology (IT) assets, but that the online tracking and categorization of resource usage based on observed communication patterns provide a much richer information base and enables faster and more accurate decisions in today's evolving e-business environment. For example, some typical workstation hardware could well be used in server role. While the maintenance of higher security and availability levels is essential for servers, this increases the IT cost. Hence early server-role detection is key to operational planning and financial compensation. In general, detailed asset configuration and usage records enable integrated Infrastructure Resource Management (IRM) processes, with value propositions ranging from network management, over help desk services, asset tracking and reporting, software distribution and license metering, service contract and warranty information, to the support of leasing, procurement, acquisition, and migration services.

Section 1 of this paper surveyed the business rationale for detailed assets inventory and monitoring in a networked world. Section 2 reviews current network-based discovery techniques, while Section 3 analyses information found in common network and application log files. Sections 4 and 5 then propose how to warehouse and compute aggregated activities to prepare for the data mining to categorize assets and users. Section 6 concludes with a validation of the log analysis techniques on a small campus network.

2 Network-Based Asset Discovery and Tracking Techniques

An increasingly popular inventory method is to collect information using the network itself. Network-based inventories can significantly reduce the time and cost of an IT audit and can also make regularly scheduled inventories feasible, providing more up-to-date information [2]. Building an accurate inventory of computing assets in an unknown systems and networking environment is a challenging task. Redundant sources of information may be desirable to build reliable algorithms and tools addressing largely heterogeneous environments. Accounting records, physical inventory baselines, end-system configurations, traffic monitors, networking services, and historic network and application server logs all represent valid sources of information (see Figure 1). The ultimate goal is to integrate all incoming records into a consistent model of what is out there and how it is being used [16].

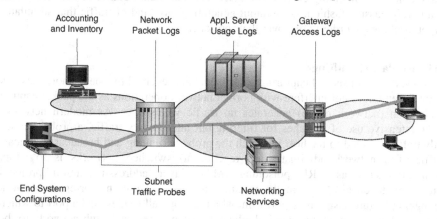

Fig. 1. Networks, Communication Paths, and Sources of Information

Network-based asset discovery and tracking techniques can be classified into "online" methods (to determine the actual state of end-systems, network and services) and "historic" log information processing (to analyze recorded network and services usage traces). Although online monitoring may also keep historic information, it cannot see into the past, i.e. into events that happened prior to its start. If the time period available for asset discovery is limited, i.e. too short to see all sporadically connecting devices, or if it is difficult to obtain sufficiently broad and deep administrative access to the overall infrastructure, it is attractive to reconstruct the global picture of computing assets and their usage from historic and redundant log information [10].

Before focusing on the analysis of log information in the subsequent chapters, the following section surveys currently known techniques for online discovery. As described in [27], there is no single technique that can perform an exhaustive network discovery, as every network has some peculiarity (e.g. some network parts are protected by firewalls, whereas others are accessible only from a few selected hosts) and also because not all networks run the same set of network protocols. In addition, network interface cards for mobile computers can easily be shared (plug-n-play) and swapped, and more and more portables are already equipped with wireless network interfaces, making these devices difficult to track and identify as they can move and change link and network addresses during their lifetime.

Network-based online discovery techniques can be further classified into methods that (i) passively listen and map the network, (ii) actively talk and walk the network and services, and (iii) interact and footprint individual hosts and target application services. The following sections highlight and position various techniques used by the authors to explore an unknown networking environment.

2.1 Passive Network Mapping

Passive network mapping enables the discovery and identification of network assets in a purely passive fashion, i.e. without generating any kind of traffic that stimulates target machines in order to discover their presence.

Network Packet Sniffing

Packet sniffing consists of capturing packets that are received by one or more network adapters, and does not interfere with normal network operations as the packet capture application (network probe) generates no traffic whatsoever. As modern networks make intensive use of switches for filtering out unnecessary traffic, a probe can see only traffic directed to the host in which the probe is running and broadcast/multicast traffic. The network administrator's reaction to switched networks is to adopt techniques such as ARP poisoning (MAC layer address-resolution protocol interaction to enable "man in the middle" attacks) and port mirroring (a network debugging feature assigning a port from which to copy all frames, and a port to which to send those frames) to avoid duplicating probes on each sub-network to be monitored. Packet capture is location dependent; hence the probe should be placed where the traffic actually flows, which can pose quite a challenge. The probe needs to have decoders for each of the protocols the network administrator is interested in. As network traffic can be quite bursty, probes must be fast enough to handle all traffic in quasi real-time and to avoid loosing track of the ongoing traffic sessions. Applications that belong to this category include [24] and [9].

Subscription to Network and Syslogs

As most of the network devices and services store activity reports in log files which can easily get enabled for incremental remote forwarding via syslog, it is quite common to subscribe to such log file events for tracking network activities. In particular, log file data can be very useful in the case of dial-in modem servers, corporate VPN gateways, mobile users, and WAP gateways. Some drawbacks of using logs are that their format is usually fixed and not customizable by network administrators, that it is necessary to periodically read the logs as they can wrap and hence overwrite historical information, and that access typically requires administrator privileges.

2.2 Active Network Mapping

There are several different techniques that can be employed for actively mapping network assets. They all share the principle that the network needs to be exhaustively explored from a starting point using a repetitive algorithm that walks the entire network up to an endpoint or until the entire IP address range has been exhausted.

SNMP Walking of Network Topology

Starting point: the current default route of the host that performs the mapping. Recursive propagation algorithm: using SNMP [1] contact all adjacent routers, learn all their current interfaces, and read their ARP table for learning all local hosts. Applications that belong to this category include [21]. Specific MIBs can also provide hardware-related configuration information, e.g. allow the algorithm to drill down to racks, frames, and individual plugs in wiring closets. Termination condition: recurs until network closure or until a limited access authorization (SNMP community string) blocks walking. The technique is potentially hazardous for networks as SNMP traffic is not suitable for large networks and can interfere with normal operations. For security reasons, network administrators may deny remote SNMP GET operations. Moreover, SNMP can be either disabled or misconfigured.

Network-Wide IP Ping Sweeps

Starting point: the host that performs the mapping. Contact all hosts of the network being explored, e.g. using ICMP ECHO (a.k.a. ping). Terminate when all addresses have been tried or when a firewall/router blocks the traffic. Evaluation: End-to-end method that works in many situations where SNMP router walking is blocked. NAT and firewall devices block inbound IP ping sweeps, whereas unmanaged IP addresses can still talk to most servers, even in other network segments. Hence the starting point for ping sweeps should be selected carefully. Ping typically works as long as end systems have an inherent business need to communicate. The generation of ICMP traffic may interfere with normal operations. ICMP ECHO can be (partially) disabled for security reasons. Other techniques [22] [31] may produce better results in terms of efficiency and accuracy.

DNS Network Domain Name-Space Walking

Starting point: the local DNS server. Algorithm: walk the DNS space by performing a zone transfer in order to know all known hosts and DNS servers. Recurs until network closure or until a DNS forbids the zone transfer. Evaluation: technique can produce misleading data as some DNS servers may be out of synchronization with the actual network state. Typically provides good information about stable network services and applications; naming conventions may allow further conclusions on intended main host usage (DNS, Notes, Mail Exchange, etc.). DNS walking is useless in non-IP networks of course, and fails on networks in which names have not been configured. Its results need to be carefully analyzed in particular when dynamic address protocols (e.g. BOOTP and DHCP) are in use.

DHCP Lease Information

Starting point: the local DHCP service or administrative access to the corresponding server. There is no standardized access across products. Microsoft's Win/NT Resource Kit contains utilities to find DHCP servers and list clients. Resulting data contains the currently assigned IP addresses with associated MAC address as key to client information. The value of this technique lies in particular in the tracking of devices connected only sporadically with the network.

Windows and Novell Network-Domains, LDAP, and Active Directory

Starting points: the local directory services of LDAP and the corresponding Microsoft and Novell application networking. This technique nicely complements DNS walking on IP-based networks. On networks in which dynamic DNS is implemented results can partially overlap owing to misconfigurations; directories tend to have richer data [18].

2.3 Host and Service Mapping

As hosts get discovered, the procedure typically continues by drilling down on each active host. We may want to employ specialized tools to learn about the currently running operating system (OS) and the services the host provides. One of the principles of network mapping is that the more we know about a host the more we can find out. The easiest way to learn the OS is to parse the logon banners a server returns when opening TCP/IP ports. Unfortunately, not all hosts offer such services. However, a large class of workstations provides further Windows-specific data. If both approaches fail, the ultimate resort is to use advanced techniques to directly analyze the TCP/IP stack.

TCP/IP Stack Analysis and OS Detection

The standardized TCP/IP protocols allow a certain degree of local-system freedom. Such local choices may impact the system tuning and application performance. Hence different stack implementations feature slightly differing behavior, especially when challenged with peculiar protocol situations such as bogus flags, out-of-scope packets, or windowing information. The current internet-shared database contains fingerprints for more than 500 IP implementations. The most popular tool is [22]. The results of stack-based OS analysis must be carefully interpreted as they are based on heuristics that can produce vague or even wrong results. Moreover, care must be exercised when interacting with some - typically old and badly maintained - hosts as the odd requests against the stack may crash the host currently being mapped.

UDP/TCP Port Scans

Port scanning is the technique that tries to communicate with remote ports, and map the TCP/IP services available from a host. Access can be tried by using either a small list of well-known ports (such as TELNET, FTP, HTTP and POP3), the entire range of named services, or by scanning the entire 64K-large port range. In addition, access can stop when granted (early close) or continue (full open) until the banners are displayed. The former can be an alternative to ping in network environments that block ICMP packets. There are also various tools available to security analysts that further scan for possible security vulnerabilities [23]. To reduce the impact on end-systems and visibility in intrusion-detection systems, "stealth" modes are commonly available to sequence randomly across IP addresses and ports. Unfortunately port scan is a potentially hostile activity, hence it needs to be used carefully and only after that the local network administrators have been informed. There is a growing list of personal tools that are able to detect port scans[2].

Remote Windows Fingerprinting
For Windows systems, there are specialized scanners that connect to the remote interfaces of Windows systems management. In particular, and especially with proper credentials, this yields a wealth of information on the hardware, networking, and software configuration of the remote host. An example is WinFingerprint [33].

In the literature and in our experience, all the techniques described above produce good results although none is accurate by itself. In fact, to improve mapping accuracy it is necessary to combine various techniques and filter out results using advanced data-mining techniques. This is the approach we use for mapping network hosts and services. The following chapter explains which techniques enabled the authors to extend the map to mobile computers and learn more about asset usage in general.

3 Processing Logs to Track Assets and Their Usage

The ability to access information virtually anywhere - at any time - is transforming the way we live and work. Pervasive computing encompasses the dramatically expanding sphere of computers from small gadgets to computers embedded within and intrinsically part of larger devices. Such devices are characterized as not always powered on, and may not always be connected to a network. Not each pervasive device is a wireless terminal. Some connected devices are not IP networked. Pervasive devices typically have a shorter life cycle than classical workstations do [15]. Together, these effects make asset management hard.

Roaming individuals may remotely access corporate resources from corporate and third-party devices, using various network access points. While this renders the borderline of an organization fuzzier, the tracking of corporate assets and resource usage remains a business-critical issue. The network discovery methods described, which typically poll a tightly connected network and systems environment, tend to fall short in what they observe in this case. It may not be feasible to subscribe to network access servers and events in real time. As in the case of tracking shy deer that visit the watering holes only at night, we may have to live with after-the-fact analysis. Even if the device has already disappeared, there is still a wealth of traces recorded at always-on servers. Analyzing server logs actually allows computers to be tracked via communication traces over extended periods in much shorter time.

---

[2] Scanlogd http://www.openwall.com/scanlogd/, and
    Norton Personal Firewall http://www.symantec.com/sabu/nis/npf/

Consider, for example, polling access-specific gateways, IP address lease records, processing packet/session logs of firewalls and VPN servers, and analyzing application-specific logs from Intranet web, mail, and other subsystems. Figure 2 shows a model of the data contained in log files from the following network services: SOCKS network proxy gateways, DNS name servers, HTTP web servers, POP/IMAP email access and SMTP mail submission servers. Despite the many differences among the various protocols and log entry formats, there are some obvious commonalities with respect to referencing external objects such as hosts, ports, and users (depicted as self-contained entities in the center of the picture).

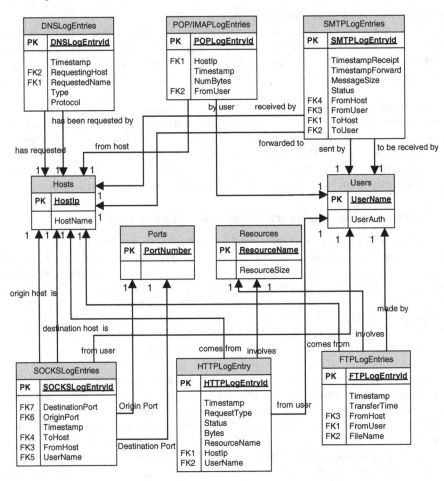

Fig. 2. Entity-Relational Modeling of Data Contained in Network Services Log Files

In summary, the growing complexity and dynamics of a typical IT environment are counterbalanced by the wealth of log information all around. While physical inventories are costly and can hardly face the dynamics of the growing number and diversity of end-user devices, substituting online methods is aligned with the trend to lean administration. This does not prevent - indeed, may require - tight control of the use of the organization's resources. We anticipate that in the long run the gathering of usage information from the various network and application subsystems is the most cost-effective asset and usage management scheme. Being the best method for pervasive devices, it may actually become standard usage also for the management of other, more static hosts, always-on workstations and servers. In the following, we focus on investigating this approach.

4 Warehousing Asset Usage Records

Again, our proposed overall approach is to complement basic network-based discovery with the combined log information from network and application servers, and then to compute an aggregate picture of assets and categorize their usage with data-mining techniques. Figure 3 depicts the stages for warehousing the log information in a relational database.

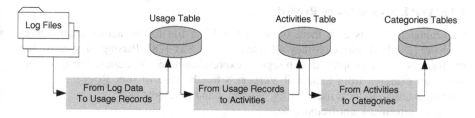

Fig. 3. Parsing, Cleaning, Coding, and Aggregating Log Information

Log files from selected network and application servers are first normalized into a denser data representation which we call usage records. This step also allows abstracting from the specific application server products and log-file formats. In a second stage (described in Section 5) related usage records from multiple network and application protocols and observation points are aggregated into a server-independent perception of activities. Usage and activity records are then input to analytic processing and data mining.

In our study we focused on log information related to services that are most widely used by network users today. We processed HTTP web and SMTP/POP/IMAP mail application logs, and analyzed TCPDUMP network packet logs. We will show how we can benefit from this redundancy of data belonging to several servers that manage different protocols. For example, if we see an HTTP activity originating from a particular host and would like to associate this with a user, we can search for email activities that originated from the same host in the same interval.

4.1 HTTP Logs and Usage Records

The W3 Consortium has standardized HTTP log file formats; in addition there is the basic Common Log File Format and the Extended Common Log File Format, which adds Referer and UserAgent information fields [32]. Fortunately, most web server products follow these formats. Deviations can be absorbed with a regular-expression-based parsing front end. The following shows an example of the original web server log file entries for accessing the root web page (index.html) on host 131.114.4.XX:

131.114.4.XX - - [25/Aug/2001:22:54:16 +0200] "GET / HTTP/1.0" 200 4234 "-"
"Mozilla/4.71 [en] (WinNT; I)"
131.114.4.XX - - [25/Aug/2001:22:54:16 +0200] "GET /images/header.gif HTTP/1.0" 200
9342 "http://www.di.unipi.it/" "Mozilla/4.71 [en] (WinNT; I)"

The corresponding consolidated single usage record stored into the database looks as follows:

ID	Record Type	StartTime	EndTime	Initiating User	Initiating Host	Target User	Target Host	Global Ref	Local Ref	Description	Data Pkts	Data Vol
1	Http Session	2001-08-25 22:54:16	2001-08-25 22:55:20		131.114.4.XX		www.di. unipi.It	/	/images/header.gif ...	Mozilla/4.71 [en] (WinNT; I)	7	45701

4.2 SMTP Logs and Usage Records

Unfortunately in this case there is no standard log file format across email server products [26] [19], although they all relate to RFC821 [28]. Parsing of specific log file formats can be simplified with regular-expression-based frontends. The following shows an example from Sendmail Version 8 for the case of user Rossi@di.unipi.it sending an email to local user Verdi@di.unipi.it and to remote user Bianchi@informatik.uni-freiburg.de.

Jun 18 09:26:37 apis sendmail[30933]: JAA14975: from=<rossi@di.unipi.it>, size=1038, class=0, pri=61038, nrcpts=2, msgid=<005101c0f7ee$54e36640$5b027283@kdd>, proto=SMTP, relay=pc-rossi [131.114.2.91]
Jun 18 09:27:06 apis sendmail[30934]: JAA14975: to=<verdi@di.unipi.it>, ctladdr=<rossi@di.unipi.it> (15124/110), delay=00:00:29, xdelay=00:00:00, mailer=local, stat=Sent
Jun 18 09:27:06 apis sendmail[30934]: JAA14975: to=<bianchi@informatik.uni-freiburg.de>, ctladdr=<rossi@di.unipi.it> (15124/110), delay=00:00:29, xdelay=00:00:28, mailer=esmtp, relay=mailgateway1.uni-freiburg.de. [132.230.1.211], stat=Sent (OK id=15BxcV-0003Xy-00)

The corresponding usage records stored into the database look as follows:

ID	Record Type	StartTime	EndTime	Initiating User	Initiating Host	Target User	Target Host	Global Ref	Local Ref	Data Pkts	Data Vol
1	MsgSending	2001-06-18 09:26:37		Rossi @di.unipi. it	pc-rossi.di.unip i.it			Msg1 @di. unipi. it	JAA1497 5	1	1038
2	LclForwarding	2001-06-18 09:27:06	2001-06-18 09:27:06			Verdi @di.unipi.it			JAA1497 5	1	
3	RmtForwarding	2001-06-18 09:27:06	2001-06-18 09:27:xx			Bianchi @informatik .uni- freiburg.de	mailgatew ay1.uni- freiburg.d e		JAA1497 5	1	

Note that we chose not to compress the data into a single record in order to maintain easy access to the individual destinations in the usage table.

4.3 POP Logs and Usage Records

Post Office Protocol Version 3 (POP) is an Internet protocol that allows a client to download email from a server to his or her Inbox on the local host, where messages are then managed. This is the most common protocol, and works well for computers that are unable to maintain a continuous connection to a server. Internet Mail Access Protocol Version 4 (IMAP) enables a client to access email on a server rather than downloading it. The following is an example where user Rossi starts a POP session from his host pc-rossi.di.unipi.it in order to download all emails that arrived since the last poll:

Jun 18 09:26:49 apis ipop3d[733352]: pop3 service init from 131.114.2.91
Jun 18 09:26:50 apis ipop3d[733352]: Auth user=Rossi host=pc-rossi.di.unipi.it [131.114.2.91] nmsgs=32/32
Jun 18 09:26:51 apis ipop3d[733352]: Logout user=Rossi host=pc-rossi.di.unipi.it [131.114.2.91] nmsgs=27 ndele=5

The corresponding usage record stored into the database looks as follows:

ID	Record Type	StartTime	EndTime	Initiating User	Initiating Host	Target User	Target Host	Global Ref	Local Ref	Data Packets	Data Volume
1	Pop Session	2001-06-18 09:26:49	2001-06-18 09:26:51	Rossi@di.unipi.it	Pc-rossi.di.unipi.it		popserver.di.unipi.it			27	32

4.4 TCPDUMP Packet Logs and Usage Records

Network packet logs definitely provide the largest volume of data with an immediate need for compression. Aggregating all entries referring to the same TCP connection into a single usage record was not sufficient (tcpdump compression tool www.tcptrace.org), especially for HTTP V1 clients that keep multiple connections to the same server open at the same time. Therefore, we decided to aggregate all data of all simultaneous connections between two hosts into a single usage record. Each such connection is considered closed when the time elapsed since receipt of the last packet is greater than a fixed gap time period. After an idle gap, further packets between the same pair of hosts result in the creation of a new usage record. We obtained best results with gap values of approx. 1 min. The following example shows a situation in which host 131.114.2.1XY has a POP session with server 217.58.130.18, followed by another host 131.114.4.1ZZ having an HTTP session with a web server on 212.48.9.22 using two simultaneous connections on ports 2099 and 2100, respectively:

999598392.171337 > 131.114.2.1XY.45316 > 217.58.130.18.pop3: S
3331168056:3331168056(0) win 5840 <mss 1460,sackOK,timestamp 364005685 0,nop,wscale 0> (DF)
...

999598421.515854 > 131.114.2.1XY.45320 > 217.58.130.18.pop3: S
3369225773:3369225773(0) win 5840 <mss 1460,sackOK,timestamp 364008620 0,nop,wscale
0> (DF)
...
999597426.543181 > 131.114.4.1ZZ.2099 > 212.48.9.22.www: S 2586282406:2586282406(0)
win 16384 <mss 1460,nop,nop,sackOK> (DF)
...
999597471.802370 > 131.114.4.1ZZ.2099 > 212.48.9.22.www: R 2586282731:2586282731(0)
win 0 (DF)

The corresponding usage record stored into the database looks as follows:

ID	Record Type	StartTime	EndTime	Initiating User	Initiating Host	Target User	Target Host	Global Ref	Local Ref	Data Packets	Data Volume
1222	Pop Session	2001-09-04 12:13:12	2001-09-04 12:13:21		131.114.2.1XY		217.58.130.18			24	236
2133	Http Session	2001-09-04 11:57:06	2001-09-04 11:57:51		131.114.4.1ZZ		212.48.9.22			21	1732

In addition to the columns shown, each usage record also contains Tmin, Tmax,
Taverage and Tstddev fields that statistically describe the time relationships between
the constituent log file entries of the usage record.

5 Computing Aggregated Activities to Prepare for Data Mining

As described above, log file entries were first consolidated into denser usage records.
As a next step, usage records originating from multiple protocols and observation
points are aggregated into a server-independent perception of activities. We continue
with our example of sending an email to a local and a remote user. Usage record 1
computed from the network packets between pc-rossi and mailserver gets combined
with usage records 2-4 derived from the mail server log.

UsageRecords

ID	Category	Server	Source	Record Type	StartTime	EndTime	Initiating User	Initiating Hostname	Target User	Target Hostname	Global Ref	Local Ref	Data Pkts	Data Vol
1	Net Log	net-gw.di. unipi.it	/etc/logs/net .log	NetSmtp Packets	2001-06-18 09:26:36			pc-rossi.di. unipi.it		mailserver.di. unipi.it			xx	1038
2	SMTP Log	mailserver.di. unipi.itt	/etc/logs/Mail y20.txt	Message Sending	2001-06-18 09:26:37		Rossi @di. unipi.it	pc-rossi.di. unipi.it			Msg1 @di.un ipi.it	JAA 149 75	1	1038
3	SMTP Log	mailserver.di. unipi.it	/etc/logs/Mail y20.txt	LocalMsg Forwarding	2001-06-18 09:27:06	2001-06-18 09:27:06			Verdi @di.unipi .it			JAA 149 75	1	
4	SMTP Log	mailserver.di. unipi.it	/etc/logs/Mail y20.txt	RemoteMsg Forwarding	2001-06-18 09:27:06	2001-06-18 09:27:07			Bianchi @inform atik.uni- freiburg.de	mailgate way1. uni- freiburg.de		JAA 149 75	1	

The result is just one activity record. Note that further usage records from other network observation points and mail servers would just confirm this single activity record with global perspective. The relationship between activities and their constituent usage records is maintained in a separate table. Auxiliary tables are also used to index the hosts and users involved in activities.

Activities

ID	Category	Server	Source	Record Type	StartTime	EndTime	Initiating User	Initiating Hostname	Target User	Target Hostname	Global Ref	Local Ref	Data Packets	Data Volume
5	activity			Email Sending	2001-06-18 09:26:36	2001-06-18 09:27:07	Rossi @di.un ipi.it	pc-rossi.di. unipi.it			Msg1 @di. unipi .it		xx+3	1038

Activity UsageRecords

ID	ActivityID	UsageRecordID
1	5	1
2	5	2
3	5	3
4	5	4

Activity Hosts

ID	ActivityID	HostName
1	5	pc-rossi.di.unipi.it
2	5	mailserver.di.unipi.it
3	5	mailgateway1.uni-freiburg.de

Activity Users

ID	ActivityID	UserName
1	5	Rossi@di.unipi.it
2	5	Verdi@di.unipi.it
3	5	Bianchi@informatik.uni-freiburg.de

With these aggregation algorithms, continuous web surfing originating at a particular host results in a single activity record. A new activity is created after an inactivity period of 60 min. An activity ends with the last end time of the constituent usage records. Concurrent web surfing and email processing result in separate activities of different types. Sending and checking/receiving email also result in separate activities of different types. When not exceeding the 1-hour inactivity period, sending multiple emails results in a single activity. A background daemon frequently checking for arriving email also results in a single activity.

6 Process Validation on a University Network

To validate the process of combining log information from network and application servers to compute an aggregate picture of computers and users according to detected communication patterns, we tested it with the Computer Science departmental staff network of the University of Pisa. This network of 512 possible IP addresses is rather heterogeneous and mostly unprotected, with the exception of a few systems used for accounting purposes that are shielded by a packet filter firewall. There exist a total of approx. 300 hosts (50 servers and 250 workstations). Under a confidentiality agreement we were able to get access to a full 7-day week of real traffic logs from the departmental web and mail servers and the gateway to the rest of the university networks. The validation playground is depicted in Figure 4.

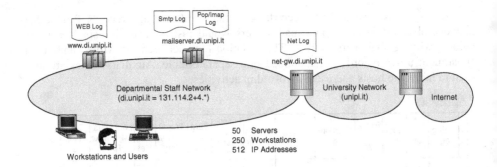

Fig. 4. Actual Network and Log-File Sources at University of Pisa

A significant problem tackled by the authors is the large amount of data to parse. The first week in September in the setup described corresponds to a total of 13 GB uncompressed raw TCPDUMP, 46 MB HTTP, and 8+7 MB POP/SMTP log files. The warehousing of usage records by means of a small Java program took several hours, parsing approx. 100,000,000 log entries, directly filtering out 30% as not-studied protocols, and creating 335,000 usage records. The usage records distribute as follows over the protocols studied: 64% HTTP, 19% SMTP, 14% POP, 3% IMAP. The distribution of the usage-record log source: 68% Net, 10% Web, 22% Mail. In other words, most usage records correspond to Internet web surfing logged by the network gateway.

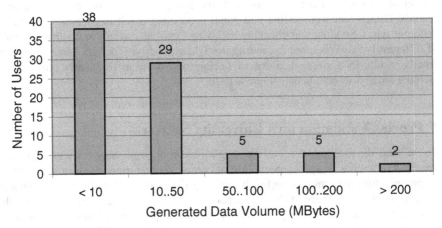

Fig. 5. Categorization of Users According to Generated Traffic

Finally, the usage records were aggregated into 6,700 activities, as consolidated input for the following data analysis.

Details of the validation work to test both the warehousing software and the concepts for categorizing hosts and users can be found in [10]. Some extracts from our findings are summarized here. The protocol used most in terms of usage records (with 1 min.

as usage gap) is definitively HTTP with 64%. As aggregated activities (with 60 min as activity gap) web surfing still represent 23%, whereas email sending and downloading activities are at 53% and 24%, respectively. 51% of the hosts analyzed are used by a single user, hence can be considered private workstations. Most of the traffic occurs during working hours (10AM – 6PM). The total number of activities increases with overall traffic, and the activity duration increases with the activity traffic. Most of the hosts that generate web traffic also send out emails. Most of the hosts generate less than 50 MB of traffic per day. Web activities usually last at least 10 min. 25 of the 79 identified users trigger email downloading explicitly; the others download emails periodically with a polling time above 10 min. For an example of a more advanced data exploration attempt, refer to Figure 5, which categorizes users with at least one personal host according to the total weekly web and email data volume generated by his/her hosts.

Eventually, we would like to derive conclusions such as: "Computer A is used by a secretarial person 5/7 days a week in the morning. Computer B probably is a student-lab workstation, shared by users X, Y and Z." The lessons learned during the validation process are that (i) there is hope to achieve this – eventually – but (ii) it is of utmost importance to minutely parse, clean, code, and compress the original data sources, and (iii) there is no way around having a baseline sample population of users and computers to establish a data-mining model allowing OLAP predictions in newly discovered environments.

7 Conclusion

In today's dynamic information society, organizations critically depend on the underlying computing infrastructure. Tracking computing devices as assets and their usage helps in the provision and maintenance of an efficient, optimized service. Building an accurate inventory of computing assets is especially difficult in unknown heterogeneous systems and networking environments without prior device instrumentation. User mobility and mobile, not-always-signed-on, computing devices add to the challenge. We therefore propose to complement basic network-based online discovery techniques with the combined historic log information from network and application servers to compute an aggregate picture of assets, and to categorize their usage with data-mining techniques according to detected communication patterns.

In this paper we outlined the process of warehousing and analyzing network and application server logs to track assets and their usage. Given our initial validation, we hope to establish the potential of integrating the consolidated historic knowledge residing in access-specific gateways, firewalls, VPN servers, and network proxies, and the growing wealth of application-specific servers. We anticipate that in the long run the gathering of usage information from the various network and application subsystems will prove to be the most cost-effective asset and usage management scheme. Having already been established as the best method for pervasive devices, it may actually become standard usage for unknown heterogeneous environments and nicely complement the management of other, more static hosts, always-on workstations and servers.

8 Future Work

The authors are aware that this work is not complete as there are open problems and challenges in categorizing computing assets. Referring to the lessons learnt during the validation process, the value of the warehoused data increases with more careful cleaning and coding. A baseline population of assets and users is then essential as further input for OLAP processing to establish a data-mining model allowing later predictions in unknown environments. In particular, this work also needs to be extended in order to categorize accurately mobile and pervasive devices that can roam across different networks and be active only for a limited period of time.

Network intrusion-detection and customer-relationship management are established fields that benefit from OLAP techniques. We hope to promote the use of similar forms of data mining also as techniques for corporate asset management and to establish a flexible and dynamic management infrastructure for e-business services. Figure 6 proposes a chain of processing steps, starting with the classical network discoveries, adding log analysis for usage categorization, that may eventually allow questions about the cost, utility, and risk associated with individual assets to be answered on the one hand, and the computation of associated values on the other.

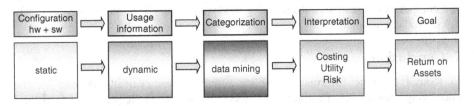

Fig. 6. Processing Stages for the Computation of Business Values

Additional ideas include the following: Associate a geographical location [20] with the categorized assets to facilitate the tasks of physically upgrading, replacing, and accounting. Enhance the proposed system with an accounting application that allows tracking the service usage, its users, and its availability. Study how to generate alarms (e.g. SNMP traps) when an asset modifies its behavior (e.g. if a computer that is known not to handle mail at some point routes emails, it means that something has changed or that a virus is running on the asset).

Acknowledgments. The proposed approach was investigated as part of a Diploma Thesis [10] and validated against real network traffic and application logs provided by the University of Pisa. In particular the authors acknowledge Prof. Maurizio Bonuccelli for his help in accessing this data, samples and statistical findings, which were also used in this paper. We would also like to acknowledge the dedication of Chris Giblin and Matthieu Verbert, who helped design and prototype most the discovery techniques described here.

References

1. J. Case et.al., Simple Network Management Protocol (SNMP), RFC 1157, 1990
2. Centennial, Network Inventory Audit, http://www.intertechnology.ca/soft-1.htm
3. R. Cooley, B. Mobasher, and J. Srivastava, Grouping Web Page References into Transactions for Mining World Wide Web Browsing Patterns, in Proc. KDEX'97, Newport Beach CA, 1997
4. M.S. Chen, J.S. Park, and P.S. Yu, Data Mining for path traversal patterns in a Web environment, in Proc. 16th Int. Conf. On Distributed Computing Systems, p. 385-392, 1996
5. Denmac Systems, Network Based Intrusion Detection: a Review of Technologies, http://www.denmac.com/, November 1999
6. L. Deri and S.Suin, Effective Traffic Measurement using ntop, IEEE Communications Magazine, May 2000, http://luca.ntop.org/ntop_IEEE.pdf.gz
7. L. Deri and S.Suin, Effective Traffic Measurement using ntop, IEEE Communications Magazine, May 2000, Monitoring Networks using Ntop, in Proc. 2001 IEEE/IFIP Int. Symp. On Integrated Network Management, Seattle, WA, 2001
8. L. Deri and S.Suin, Ntop: beyond Ping and Traceroute, DSOM, 1999, http://luca.ntop.org/ntop_DSOM99.pdf.gz
9. Ethereal free network protocol analyzer for Unix and Windows, http://www.ethereal.com/
10. M. Filoni, Computing assets categorization according to collected configuration and usage information, Diploma Thesis, University of Pisa, Italy, November 2001
11. D. Gantenbein, Network-based IT asset discovery and categorization, Presentation, University of Pisa, Italy, October 2000
12. J. Han and M. Kamber, Data Mining: Concepts and Techniques, Academic Press, 2001, ISBN 1-55860-489-8
13. D. Heywood, Networking with Microsoft TCP/IP, 3rd edition, Microsoft Press, New Riders Publishing, ISBN: 0-7357-0014-1, 1998
14. J. Fielding et.al., Hypertext Transfer Protocol – HTTP 1.1
15. Pervasive Computing, IBM Systems Journal, Vol. 38, No. 4, 1999. http://www.research.ibm.com/journal/sj38-4.html
16. Intelligent Device Discovery (IDD), IBM Zurich Research Laboratory, Project http://www.zurich.ibm.com/csc/ibi/idd.html
17. V. Jacobson, C. Leres, and S. McCanne, *tcpdump*, Lawrence Berkeley National Labs, ftp://ftp.ee.lbl.gov/, 1989
18. A. G. Lowe-Norris, Windows 2000 Active Directory, O'Reilly, ISBN 3-89721-171-8, 2001
19. Microsoft Exchange: Tracking Log, http://www.microsoft.com/Exchange/en/55/help/default.asp?url=/Exchange/en/55/help/documents/server/XMT04027.HTM
20. Caida, NetGeo: the Internet Geographic Database, http://netgeo.caida.org/, 2000
21. NetView, Tivoli product, http://www.tivoli.com/products/index/netview/
22. NMAP Free Security Scanner, http://www.insecure.org/nmap/index.html
23. NSA Firewall Network Security Auditor, http://www.research.ibm.com/gsal/gsal-watson.html
24. L. Deri, Ntop: a Lightweight Open-Source Network IDS, 1998-2001, http://www.ntop.org/
25. J. Pitkow, In search of a reliable usage data on the www, in Sixth Int. Wold Wide Web Conf., pp. 451-463, Santa Clara CA, 1997
26. Sendmail mail service, http://www.sendmail.org
27. R. Siamwalla et al., Discovering Internet Topology, in Proc. IEEE INFOCOM '99, 1999, http://www.cs.cornell.edu/skeshav/papers/discovery.pdf
28. Simple Mail Transfer Protocol, J. B. Postel, RFC 821, August 1982

29. Tivoli, Inventory management, http://www.tivoli.com/products/index/inventory/
30. Tivoli, Performance solutions,
    http://www.tivoli.com/products/solutions/availability/news.html
31. S. Branigan et al., What can you do with Traceroute?,
    http://www.computer.org/internet/v5n5/index.htm
32. Extended Common Log File Format, W3C working Draft, http://www.w3.org/TR/WD-logfile
33. Winfingerprint Windows Information Gathering Tool,
    http://winfingerprint.sourceforge.net/

Remarks on Ad Hoc Networking

Stefano Basagni

Department of Electrical and Computer Engineering
Northeastern University
basagni@ece.neu.edu

Abstract. This papers describes selected problems and solutions for *ad hoc networking*, namely, for networking in absence of a fixed infrastructure. *All* nodes of an ad hoc networks move freely and communicate with each other only if they are in each other transmission range (neighboring nodes). This implies that in case two nodes are not neighbors, in order for them to communicate they have to rely on the forwarding services of intermediate nodes, i.e., each node is a router and the communication proceeds in multi-hop fashion. In this paper we are concerned with three aspects of ad hoc networking. The problem of accessing the wireless channel, i.e., the problem of devising *Media Access Control (MAC)* protocols. The problem of grouping the nodes of the network so to obtain a hierarchical network structure (*clustering*). The problem of setting up an ad hoc network of *Bluetooth devices*, i.e., of forming a Bluetooth *scatternet*.

1 Introduction

The ability to access and exchange information virtually anywhere, at any time, is transforming the way we live and work. Small, handheld unthethered devices are nowadays at anybody's reach, thus allowing new forms of distributed and collaborative computation. Among the several examples of this new form of communication we can mention what is often referred to as *pervasive computing*. The essence of pervasive computing is the creation of environments saturated with computing and wireless communication, yet gracefully integrated with human users. Numerous, casually accessible, often invisible computing devices, which are frequently mobile or embedded in the environment, are connected to an increasingly ubiquitus network structure.

The possible network architectures that enable pervasive computing fall into two main categories: *Cellular networks* and *multihop wireless networks* or, as commonly termed recently, *ad hoc networks*. In the first case, some specialized nodes, called *base stations*, coordinate and control all transmissions within their coverage area (or *cell*). The base station grants access to the wireless channels in response to service requests received by the mobile nodes currently in its cell. Thus the nodes simply follow the instruction of the base station: For this reason, the mobile nodes of a cellular network need limited sophistication and can request and achieve all the information they need via the base station that is currently serving them.

E. Gregori et al. (Eds.): Networking 2002 Tutorials, LNCS 2497, pp. 101–123, 2002.

The primary characteristic of an *ad hoc network architecture* is the absence of any predefined structure. Service coverage and network connectivity is defined solely by node proximity and the prevailing RF propagation characteristics. Ad hoc nodes directly communicate with one another in a peer-to-peer fashion. To facilitate communication between distant nodes, each ad hoc node also acts as a router, storing and forwarding packets on behalf of other nodes. The result is a generalized wireless network that can be rapidly deployed and dynamically reconfigured to provide on-demand networking solutions.

In this work we are concerned about this second, more general, kind of network architecture, which is recently gaining more and more attention—both from the academia and industry.

While the generic architecture of ad hoc network certainly has its advantages, it also introduces several new challenges. All network control and protocols must be distributed. For any possible collaborative task, each ad hoc node must be aware of what is happening around them, and cooperate with other nodes in order to realize critical network services, which are instead realized by the base stations in a cellular environment. Considering that most ad hoc systems are fully mobile, i.e., each node moves independently, the level of protocol sophistication and node complexity is high. Power conservation is also of the essence, since most of the devices of upcoming ad hoc networks, such as handheld devices, laptops, small robots, sensors and actuators are battery operated. Finally, networks operations and protocols should be *scalable*, i.e., largely independent of the increasing number of networks nodes, or of their larger geographical distribution.

In this paper we describe some results that have been proposed in recent years on ad hoc networking. In particular here we focus on two main aspects, namely, *Media Access Control (MAC) protocols*, i.e., methods for successfully accessing the wireless channel, and *clustering* protocols, i.e., protocols for the grouping of nodes into clusters. As an application of clustering, we illustrate a protocol for the set up of an ad hoc network of *Bluetooth* devices, Bluetooth being a wireless technology that enables ad hoc networking. The important issue of routing and in general, of multipoint communication, are not dealt with in this paper. These issues have been widely covered in many comprehensive survey papers, such as [1] and [2].

The rest of the paper is organized as follows. In the next section we describe the problem of accessing the wireless channel, and the proposed MAC protocols that solve this problem. In Section 3 we give an example of a clustering algorithm which is well suited for mobile ad hoc networks. We finally illustrate an application of the presented clustering algorithm for setting up ad hoc networks of Bluetooth devices (Section 4). Section 5 concludes the paper.

2   Ad Hoc MAC Protocols

Ad hoc networks do not have the benefit of having predefined base stations to coordinate channel access, thus invalidating many of the assumptions held by MAC designs for the centralized (cellular) architecture. In this section, we focus our attention on MAC protocols that are specifically designed for ad hoc networks. (This section is based on the research performed with Dr. Andrew Myers, and can be more thoroughly found in [3]).

We start by exploring the physical constraints of the wireless channel and discuss their impact on MAC protocol design and performance.

Radio waves propagate through an unguided medium that has no absolute or observable boundaries and is vulnerable to external interference. The signal strength of a radio transmission rapidly attenuates as it progresses away from the transmitter. This means that the ability to detect and receive transmissions is dependent on the distance between the transmitter and receiver. Only nodes that lie within a specific radius (the *transmission range*) of a transmitting node can detect the signal (carrier) on the channel. This location dependent carrier sensing can give rise to so-called *hidden* and *exposed* nodes that can detrimentally affect channel efficiency. A hidden node is one that is within range of a receiver but not the transmitter, while the contrary holds true for an exposed node. Hidden nodes increase the probability of *collision* at a receiver, while exposed nodes may be denied channel access unnecessarily, thereby under utilizing the bandwidth resources.

Performance is also affected by the signal propagation delay, i.e., the amount of time needed for the transmission to reach the receiver. Protocols that rely on carrier sensing are especially sensitive to the propagation delay. With a significant propagation delay, a node may initially detect no active transmissions when, in fact, the signal has simply failed to reach it in time. Under these conditions, collisions are much more likely to occur and system performance suffers. In addition, wireless systems that use a synchronous communications model must increase the size of each time slot to accommodate propagation delay. This added overhead reduces the amount of bandwidth available for information transmission.

A possible taxonomy of ad hoc MAC protocols includes three broad protocol categories that differ in their channel access strategy: *Contention protocols*, *allocation protocols*, and a combination of the two (*hybrid protocols*).

Contention protocols use direct competition to determine channel access rights, and resolve collisions through randomized retransmissions. Prime examples of this protocols are ALOHA and CSMA (for a brief discussion on "core" MAC protocols such as ALOHA, slotted ALOHA, CSMA, TDMA, FDMA and CDMA the reader is referred to [3]). With the exception of slotted ALOHA, most contention protocols employ an asynchronous communication model. Collision avoidance is also a key design element that is realized through some form of control signaling.

The contention protocols are simple and tend to perform well at low traffic loads, i.e., when there are few collision, leading to high channel utilization and low packet delay. However, protocol performance tends to degrade as the traffic loads are increased and the number of collisions rise. At very high traffic loads, a contention protocol can become unstable as the channel utilization drops. This can result in exponentially growing packet delay and network service breakdown since few, if any, packets can be successfully exchanged.

Allocation protocols employ a synchronous communication model, and use a scheduling algorithm that generates a mapping of time slots to nodes. This mapping results in a transmission schedule that determines in which particular slots a node is allowed to access the channel. Most allocation protocols create collision-free transmission schedules, thus the schedule length (measured in slots) forms the basis of protocol performance. The time slots can either be allocated statically or dynamically, leading to a fixed and variable schedule length.

The allocation protocols tend to perform well at moderate to heavy traffic loads as all slots are likely to be utilized. These protocols also remain stable even when the traffic loads are extremely high. This is due to the fact that most allocation protocols ensure that each node has collision-free access to at least one time slot per frame. On the other hand, these protocols are disadvantaged at low traffic loads due to the artificial delay induced by the slotted channel. This results in significantly higher packet delays with respect to the contention protocols.

Hybrid protocols can be loosely described as any combination of two or more protocols. However, in this section, the definition of the term hybrid will be constrained to include only those protocols that combine elements of contention and allocation based channel access schemes in such a way as to maintain their individual advantages while avoiding their drawbacks. Thus the performance of a hybrid protocol should approximate a contention protocol when traffic is light, and an allocation protocol during periods of high load. (For details on hybrid protocols the reader is referred to [3].)

2.1   Contention Protocols

Contention protocols can be further classified according to the type collision avoidance mechanism employed. The ALOHA protocols make up the category of protocols that feature no collision avoidance mechanism, i.e., they simply react to collision via randomized retransmissions. Most contention protocols, however, use some form of collision avoidance mechanism.

The busy-tone multiple access (BTMA) protocol [4] divides the entire bandwidth into two separate channels. The main *data channel* is used for the transmission of packets, and occupies the majority of the bandwidth. The *control channel* is used for the transmission of a special *busy-tone signal* that indicates the presence of activity on the data channel. These signals are not bandwidth intensive, thus the control channel is relatively small.

The BTMA protocol operates as follows. When a source node has a packet to transmit, it first listens for the busy-tone signal on the control channel. If the control channel is idle, i.e., no busy-tone is detected, then the node may begin transmitting its packet. Otherwise, the node reschedules the packet for transmission at some later time. Any node that detects activity on the data channel immediately begins transmitting the busy-tone on the control channel. This continues until the activity on the data channel ceases.

In this way, BTMA prevents all nodes that are two hops away from an active source node from accessing the data channel. This significantly lowers the level of hidden node interference, and therefore reduces the probability of collision. However, the number of exposed nodes is dramatically increased. The consequence being a severely underutilized data channel.

The receiver initiated busy-tone multiple access (RI-BTMA) protocol [5] attempts to minimize the number of exposed nodes by having only the destination(s) transmit the busy-tone. Rather than immediately transmitting the busy-tone upon detection of an active data channel, a node monitors the incoming data transmission to determine whether it is a destination. This determination takes a significant amount of time, especially in a noisy environment with corrupted information. During this time, the initial transmission remains vulnerable to collision. This can be particularly troublesome in high speed systems where the packet transmission time may be short.

The wireless collision detect (WCD) protocol [6] essentially combines the BTMA and RI-BTMA protocols by using two distinct busy-tone signals on the control channel. WCD acts like BTMA when activity is first detected on the main channel, i.e., it transmits a *collision detect* (CD) signal on the BTC. RI-BTMA behavior takes over once a node determines it is a destination. In this case, a destination stops transmitting the CD signal, and begins transmitting a *feedback-tone* (FT) signal. In this way, WCD minimizes the exposed nodes while still protecting the transmission from hidden node interference.

These busy-tone protocols feature simple designs that require only a minimal increase in hardware complexity. Because of its unique characteristics, the WCD protocol is the overall performance leader followed by RI-BTMA and BTMA, respectively [6]. Furthermore, the performance of busy-tone protocols are less sensitive to the hardware switching time since it is assumed that a node can transmit and receive on the data and control channels simultaneously. However, wireless systems that have a limited amount of RF spectrum may not be able to realize a separate control and data channel. In such cases, collision avoidance using in-band signaling is necessary.

The multiple access with collision avoidance (MACA) protocol [7] uses a handshaking dialogue to alleviate hidden node interference and minimize the number of exposed nodes. This handshake consists of a *request-to-send* (RTS) control packet that is sent from a source node to its destination. The destination replies with a *clear-to-send* (CTS) control packet, thus completing the handshake. A CTS response allows the source node to transmit its packet. The absence of a CTS forces a node to reschedule the packet for transmission at some later time.

Notice that a hidden node is likely to overhear the CTS packet sent by a destination node, while an exposed node is not. Thus by including the time needed to receive a CTS and packet in the respective RTS and CTS packets, we reduce the likelihood of hidden node interference and the number of exposed nodes simultaneously.

The MACAW protocol [8] enhances MACA by including carrier sensing to avoid collisions among RTS packets, and a positive acknowledgement (ACK) to aid in the rapid recovery of lost packets. To protect the ACK from collision, a source node transmits a *data sending* (DS) control packet to alert exposed nodes of its impending arrival. Improvements are also made to the collision resolution algorithm to ensure a more equitable sharing of the channel resources.

The MACA with piggyback reservations (MACA/PR) protocol [9] enhances MACA by incorporating channel reservations. This allows the system to support QoS sensitive applications. Each node maintains a *reservation table* (RT) that is used to record the channel reservations made by neighboring nodes. A source node makes a reservation by first completing a RTS/CTS exchange. It then sends the first real-time packet whose header contains the time interval specifying the interval in which the next one will be sent. The destination responds with an ACK carrying the equivalent time interval. Other nodes within range note this reservation in their RT, and remain silent during the subsequent time intervals. Thus the source node can send subsequent real-time packets without contention. To ensure proper bookkeeping, the nodes periodically exchange their RTs.

The MACA by invitation (MACA-BI) protocol [10] reverses the handshaking dialogue of MACA. In this case, the destination node initiates packet transmission by sending a *request-to-receive* (RTR) control packet to the source node. The source node responds to this poll with a packet transmission. Thus each node must somehow predict when neighbors have packets for it. This requires each node must maintain a list of its neighbors along with their traffic characteristics. In order to prevent collision, the nodes must also synchronize their polling mechanisms by sharing this information with their neighbors.

These MACA based contention protocols minimize collisions by reducing the negative effect of hidden and exposed nodes through simple handshaking dialogues. However, the exchange of multiple control packets for each data packet magnifies the impact of signal propagation delay and hardware switching time. To some extent the MACA/PR and MACA/BI protocols alleviate these problems reducing the amount of handshaking, yet the amount of state information maintained at each node can be substantial.

2.2   Allocation Protocols

There are two distinct classes of allocation protocols that differ in the way the transmission schedules are computed. *Static allocation protocols* use a centralized scheduling algorithm that statically assigns a fixed transmission schedule to each node prior to its operation. This type of scheduling is equivalent to the assignment of MAC addresses for Ethernet interface cards. *Dynamic allocation protocols* uses a distributed scheduling algorithm that computes transmission schedule in an on-demand fashion.

Since the transmission schedules are assigned beforehand, the scheduling algorithm of a static allocation protocols requires global system parameters as input. The classic TDMA protocol builds its schedules according to the maximum number of nodes in the network. For a network of N nodes, the protocol uses a frame length of N slots and assigns each node one unique time slot. Since each node has exclusive access to one slot per frame, there is no threat of collision for any packet type (i.e., unicast or multicast). Moreover, the channel access delay is bounded by the frame length. Because of the equivalence between system size and frame length, classic TDMA performs poorly in large scale networks.

The time spread multiple access (TSMA) protocol [11] relaxes some of the strict requirements of classic TDMA to achieve better performance while still providing bounded access delay. The TSMA scheduling algorithm assigns each node multiple slots in a single frame, and permits a limited amount of collisions to occur. These two relaxations allow TSMA to obtain transmission schedules whose length scales *logarithmically with respect to the number of nodes*. Furthermore, TSMA guarantees the existence of a collision-free transmission slot to each neighbor within a single frame.

The source of this "magic" is the scheduling algorithm that makes use of the mathematical properties of finite fields. An excellent introduction to finite fields can be found in [12]. The scheduling algorithm is briefly outlined as follows. For a network of N nodes, the parameters q (of the form $q = p^m$, where p is a prime and m an integer) and integer k are chosen such that $q^{k+1} \geq N$ and $q \geq kD_{max} + 1$, where D_{max} is the maximum node degree. Each node can then be assigned a unique polynomial f over the Galois field $GF(q)$. Using this polynomial, a unique TSMA transmission schedule is computed where bit $i = 1$ if $(i \bmod q) = f(\lfloor i/q \rfloor)$, otherwise $i = 0$.

As shown in [11], that this TSMA scheduling algorithm provides each node with a transmission schedule with guaranteed access in each time frame. The maximum length of this schedule is bounded by:

$$L = O \left(\frac{D_{max}^2 \log^2 N}{\log^2 D_{max}} \right).$$

Notice that the frame length scales logarithmically with the number of nodes and quadratically with the maximum degree. For ad hoc networks consisting of thousands of nodes with a sparse topology (i.e., small D_{max}), TSMA can yield transmission schedules that are much shorter than TDMA. Table 1 compares the frame lengths of TDMA and TSMA for a network of $N = 1000$ nodes. For TSMA protocols a $\Omega(\log n)$ lower bound has been proved for L in [13]. We notice that there is still a gap between the TSMA upper bound and the mentioned logarithmic lower bound. Therefore, there is still room for improvements (more likely on the lower bound side). Protocols TSMA-like have also been deployed as a basis for implementing *broadcast* (i.e., one-to-all communication) in ad hoc networks. Upper and lower bound for deterministic and distributed TSMA-based broadcast can be found in [14,15] and [16], respectively.

With mobile ad hoc networks, nodes may be activated and deactivated without warning, and unrestricted mobility yields a variable network topologies. Consequently, global parameters, such as node population and maximum degree, are typically unavailable or difficult to predict. For this reason, protocols that use only local parameters have been developed. A local parameter refers to information that is specific to a limited region of the network, such as the number of

Table 1. Frame lengths of classic TDMA vs. TSMA.

	$D = 2$	$D = 5$	$D = 10$	$D = 15$
TDMA	1000	1000	1000	1000
TSMA	49	121	529	961

nodes within x hops of a reference node (referred to as an x-hop neighborhood). A dynamic allocation protocol then uses these local parameters to deterministically assign transmission slots to nodes. Because local parameters are likely to vary over time, the scheduling algorithm operates in a distributed fashion and is periodically executed to adapt to network variations.

Dynamic allocation protocols typically operate in two phases. Phase one consists of a set of reservation slots in which the nodes contend for access to the subsequent transmission slots. Lacking a coordinating base station, contention in this phase requires the cooperation of each individual node to determine and verify the outcome. Successful contention in phase one grants a node access to one or more transmission slots of phase two, in which packets are sent.

A great number of dynamic allocation protocols have been proposed. The protocols in [17]-[18] are just a few excellent examples of this two-phase design. The protocols in [17]-[19] use a contention mechanism that is based on classic TDMA. Essentially the nodes take turns contending for slot reservations, with the earliest node succeeding. This results in a high degree of unfairness which is equalized by means of a reordering policy. Although these protocols create transmission schedules that are specific to the local network topology, they still require global parameters.

In contrast, the five phase reservation protocol (FPRP) [18] is designed to be arbitrarily scalable, i.e., independent of the global network size. FPRP uses a complex frame structure that consists of two subframe types, namely *reservation frames* and *information frames*. A reservation frame precedes a sequence of k information frames. Each reservation frame consists of ℓ *reservation slots* that correspond to the ℓ *information slots* of each information frame. Thus, if a node wants to reserve a specific information slot, it contends in the corresponding reservation slot. At the end of the reservation frame, a TDMA schedule is created and used in the following k information frames. The schedule is then recomputed in the next reservation frame.

In order to accommodate contention, each reservation slot consists of m *reservation cycles* that contain a five round reservation dialogue. A reservation is made in the first four rounds, while the fifth round is mainly used for performance optimization. The contention is summarized as follows. A node that wishes to make a reservation sends out a request using p-persistent slotted ALOHA (round 1), and feedback is provided by the neighboring nodes (round 2). A successful request, i.e., one that did not involve a collision, allows a node to reserve the slot (round 3). All nodes within two hops of the source node are then notified of the reservation (round 4). These nodes will honor the reservation and make no further attempts to contend for the slot. Any unsuccessful reservation attempts are resolved through a pseudo-Bayesian resolution algorithm that randomizes the next reservation attempt.

In [18], FPRP is shown to yield transmission schedules that are collision-free, however the protocol requires a significant amount of overhead. Each reservation cycle requires a number of hardware switches between transmitting and receiving modes. Each round of contention must also be large enough to accommodate the signal, propagation delay and physical layer overhead (e.g., synchronization and guard time). Add this together and multiply the result by m reservation cycles and ℓ reservation slots, and the end result is anything but trivial. Furthermore, the system parameters k, ℓ and m are heuristically determined through simulation and then fixed in the network. This limits the ability of FPRP to dynamically adapt its operation to suit the current network conditions which may deviate from the simulated environment.

3   Clustering for Ad Hoc Networks

Among the many ways to cope with the barriers and challenges posed by the ad hoc network architecture, here we describe a possible solution based on *grouping* the network nodes into *clusters*. This operation goes commonly under the name of *clustering*.

In this section, we describe a protocol for clustering set up and clustering maintenance in presence of node mobility. The cluster are characterized by a node that coordinates the clustering process (a *clusterhead*) and possibly few non-clusterhead nodes, that have direct access to only one clusterhead (one hop, non-overlapping clusters).

In the following description of the clustering protocol we consider an *ad hoc* network as an undirected graph $G = (V, E)$ in which V, $|V| = n$, is the set of (wireless) nodes and there is an edge $\{u, v\} \in E$ if and only if u and v can mutually receive each others' transmission. In this case we say that u and v are neighbors. The set of the neighbors of a node $v \in V$ will be denoted by $\Gamma(v)$. Due to mobility, the graph can change in time.

Every node v in the network is assigned a unique identifier (ID). For simplicity, here we identify each node with its ID and we denote both with v. Finally, we consider weighted networks, i.e., a weight w_v (a real number ≥ 0) is assigned to each node $v \in V$ of the network. For the sake of simplicity, here we stipulate that each node has a different weight. (In case two nodes have the same weight, the tie can be broken arbitrarily, e.g., by using the nodes' ID.)

In this section, clustering an ad hoc network means partitioning its nodes into *clusters*, each one with a *clusterhead* and (possibly) some *ordinary nodes*. (Clusterhead and ordinary node ar the *roles* that each node may assume.) The choice of the clusterheads is here based on the *weight* associated to each node: the bigger the weight of a node, the better that node for the role of clusterhead. Each node constantly computes its weight based on what is most critical to that node for the specific network application (e.g., node mobility, its remaining battery life, and its *connectivity degree*, i.e., the number of its neighbors). For instance, as introduced in [20,21,22], we assume the following expression for the computation of node v's weight:

$$w_v = \sum_{i \in I} c_i P_i$$

where the c_is are the (constant) weighing factors for the $|I|$ system parameters of interest P_i.

The protocol described in this section is a generalization of the Distributed Mobility-Adaptive Clustering (DMAC) originally presented in [23]. DMAC is a distributed algorithm for clustering set-up and maintenance in presence of node mobility that partition the nodes of the network into "one hop" clusters in such a way that no two clusterheads can be neighbors and so that whenever a "better clusterhead" moves into the neighborhood of an ordinary nodes, the ordinary node must affiliate to the new clusterhead. Here we relax these conditions allowing clusterheads to be neighbors and allowing ordinary nodes to choose whether to switch to a new neighboring clusterhead or not.

The process of cluster formation/maintenance is continuously executed at each node, and each node decides its own role so that the following three requirements (that we call "ad hoc clustering properties") are satisfied:

1. Every ordinary node always affiliates with only one clusterhead.
2. For every ordinary node v there is no clusterhead $u \in \Gamma(v)$ such that $w_u > w_{Clusterhead} + h$, where *Clusterhead* indicates the current clusterhead of v.
3. A clusterhead cannot have more than k neighboring clusterheads.

Requirement number 1. ensures that each ordinary node has direct access to at least one clusterhead (the one of the cluster to which it belongs), thus allowing fast intra- and inter-cluster communications. This is the property that insures that this protocol is a "single hop" kind of clustering protocol. Also, since an ordinary node affiliates only to one clusterhead, the obtained clusters are not overlapping. The second requirement guarantees that each ordinary node always stays with a clusterhead that can give it a "guaranteed good" service. By varying the threshold parameter h it is possible to reduce the communication overhead associated to the passage of an ordinary node from its current clusterhead to a new neighboring one when it is not necessary. Finally, requirement number 3. allows us to have the number of clusterheads that can be neighbors as a parameter of the algorithm. This, as seen for requirement number 2. allows us to consistently reduce the communication overhead due to the change of role of nodes.

The following description of the algorithm is based on the following two common assumptions:

- A message sent by a node is received correctly within a finite time (a *step*) by all its neighbors.
- Each node knows its own ID, its weight, its role (if it has already decided its role) and the ID, the weight and the role of all its neighbors (if they have already decided their role). When a node has not yet decided what its role is going to be, it is considered as an ordinary node.

The algorithm is executed at each node in such a way that at a certain time a node v decides (to change) its role. This decision is entirely based on the decision of the nodes $u \in \Gamma(v)$ such that $w_u > w_v$.

Except for the initial procedure, the algorithm is message driven: a specific procedure will be executed at a node depending on the reception of the corresponding message. We use three types of messages that are exchanged among the nodes: $\text{CH}(v)$, used by a node $v \in V$ to make its neighbors aware that it is going to be a clusterhead, $\text{JOIN}(v, u)$, with which a node v communicates to its neighbors that it will be part of the cluster whose clusterhead is node $u \in \Gamma(v)$, $v, u \in V$, and $\text{RESIGN}(w)$ that notifies a clusterhead whose weight is $\leq w$ that it has to resign its role. In the following discussion and in the procedures we use the following notation:

- v, the generic node executing the algorithm (from now on we will assume that v encodes not only the node's ID but also its weight w_v);
- $Cluster(v)$, the set of nodes in v's cluster. It is initialized to $\emptyset$, and it is updated only if v is a clusterhead;
- $Clusterhead$, the variable in which every node records the (ID of the) clusterhead that it joins. It is initialized to **nil**;
- $Ch(-)$, boolean variables. Node v sets $Ch(u)$, $u \in \{v\} \cup \Gamma(v)$, to **true** when either it sends a $\text{CH}(v)$ message ($v = u$) or it receives a $\text{CH}(u)$ message from u ($u \neq v, u \in \Gamma(v)$).

Furthermore:

- Every node is made aware of the failure of a link, or of the presence of a new link by a service of a lower level (this will trigger the execution of the corresponding procedure);
- The procedures of the algorithm (M-procedures, for short) are "atomic," i.e., they are not interruptible;
- At clustering set up or when a node is added to the network its variables $Clusterhead$, $Ch(-)$, and $Cluster(-)$ are initialized to **nil, false** and $\emptyset$, respectively.

The following two rules define how the nodes assume/change their roles adapting to changes in the network topology.

1. Each time a node v moves into the neighborhood of a clusterhead u with a bigger weight, node v switches to u's cluster only if $w_u > w_{Clusterhead} + h$, where $Clusterhead$ is the clusterhead of v (it can be v itself) and h is a real number ≥ 0. This should happen independently of the current role of v. With this rule we want to model the fact that we incur the switching overhead only when it is really convenient. When $h = 0$ we simply obtain that each ordinary nodes affiliates to the neighboring clusterhead with the biggest weight.
2. We allow a clusterhead v to have up to k neighboring clusterheads, $0 \leq k < n$. We call this condition the k-neighborhood condition. Choosing $k = 0$ we obtain that no two clusterheads can be neighbors (maximum degree of independence: In graph-theoretic terms, the resulting set of clusterhead is an *independent set*).

The parameters h and k can be different from node to node, and they can vary in time. This allows the algorithm to self-configure dynamically in order to meet the specific needs of upper layer applications/protocols that requires an underlying clustering organization. At the same time, different values of h and k allow our algorithm to take into account dynamically changing network conditions, such as the network connectivity (related to the average nodal degree, i.e., to the average number of the neighbors of the nodes), variations in the rate of the mobility of the nodes, etc.

The following is the description of the six M-procedures.

• *Init.* At the clustering set up, or when a node v is added to the network, it executes the procedure *Init* in order to determine its own role. If among its neighbors there is at least a clusterhead with bigger weight, then v will join it. Otherwise it will be a clusterhead. In this case, the new clusterhead v has to check the number of its neighbors that are already clusterheads. If they exceed k, then a RESIGN message is also transmitted, bearing the weight of the first clusterhead (namely, the one with the $(k+1)$th biggest weight) that violates the k-neighborhood condition (this weight is determined by the operator $\min_k$). On receiving a message RESIGN(w), every clusterhead u such that $w_u \leq w$ will resign. Notice that a neighbor with a bigger weight that has not decided its role yet (this may happen at the clustering set up, or when two or more nodes are added to the network at the same time), will eventually send a message (every node executes the *Init* procedure). If this message is a CH message, then v could possibly resign (after receiving the corresponding RESIGN message) or affiliate with the new clusterhead.

PROCEDURE *Init*;
begin
    if $\{z \in \Gamma(v) : w_z > w_v \wedge Ch(z)\} \neq \emptyset$
        then begin
                $x := \max_{w_z > w_v}\{z : Ch(z)\}$;
                send JOIN(v,x);
                Clusterhead $:= x$
            end
        else begin
                send CH(v);
                $Ch(v) := $ **true**;
                Clusterhead $:= v$;
                Cluster$(v) := \{v\}$;
                if $|\{z \in \Gamma(v) : Ch(z)\}| > k$ **then**
                      send RESIGN($\min_k\{w_z : z \in \Gamma(v) \wedge Ch(z)\}$)
        end
    end;

• *Link_failure.* Whenever made aware of the failure of the link with a node u, node v checks if its own role is clusterhead and if u used to belong to its cluster. If this is the case, v removes u from *Cluster*(v). If v is an ordinary node, and u was its own clusterhead, then it is necessary to determine a new role for v. To this aim, v checks if there exists at least a clusterhead $z \in \Gamma(v)$ such that

$w_z > w_v$. If this is the case, then v joins the clusterhead with the bigger weight, otherwise it becomes a clusterhead. As in the case of the *Init* procedure, a test on the number of the neighboring clusterheads is now needed, with the possible resigning of some of them.

PROCEDURE *Link_failure*(u);
begin
    if $Ch(v)$ **and** $(u \in Cluster(v))$
        then $Cluster(v) := Cluster(v) \setminus \{u\}$
        else if $Clusterhead = u$ **then**
            if $\{z \in \Gamma(v) : w_z > w_v \wedge Ch(z)\} \neq \emptyset$
            then begin
                $x := \max_{w_z > w_v}\{z : Ch(z)\};$
                send JOIN(v,x);
                $Clusterhead := x$
            end
            else begin
                send CH(v);
                $Ch(v) := \mathbf{true};$
                $Clusterhead := v;$
                $Cluster(v) := \{v\};$
                if $|\{z \in \Gamma(v) : Ch(z)\}| > k$ **then**
                    send RESIGN($\min_k\{w_z : z \in \Gamma(v) \wedge Ch(z)\}$)
            end
end;

- *New_link*. When node v is made aware of the presence of a new neighbor u, it checks if u is a clusterhead. If this is the case, and if w_u is bigger than the weight of v's current clusterhead plus the threshold h, than, independently of its own role, v affiliates with u. Otherwise, if v itself is a clusterhead, and the number of its current neighboring clusterheads is $> k$ then the operator $\min_k$ is used to determine the weight of the clusterhead x that violates the k-neighborhood condition. If $w_v > w_x$ then node x has to resign, otherwise, if no clusterhead x exists with a weight smaller than v's weight, v can no longer be a clusterhead, and it will join the neighboring clusterhead with the biggest weight.

PROCEDURE *New_link*(u);
begin
    if $Ch(u)$ **then**
        if $(w_u > w_{Clusterhead} + h)$
            then begin
                send JOIN(v,u);
                $Clusterhead := u;$
                if $Ch(v)$ **then** $Ch(v) := \mathbf{false}$
            end
            else if $Ch(v)$ **and** $|\{z \in \Gamma(v) : Ch(z)\}| > k$ **then**
                begin
                  $w := \min_k\{w_z : z \in \Gamma(v) \wedge Ch(z)\};$
                  if $w_v > w$ **then send** RESIGN(w)

$$\textbf{else begin}$$
$$x := \max_{w_z > w_v}\{z : Ch(z)\};$$
$$\textbf{send } \text{JOIN}(v,x);$$
$$Clusterhead := x;$$
$$Ch(v) := \textbf{false}$$
$$\textbf{end}$$

                  end

  end;

- *On receiving* CH(u). When a neighbor u becomes a clusterhead, on receiving the corresponding CH message, node v checks if it has to affiliate with u, i.e., it checks whether w_u is bigger than the weight of v's clusterhead plus the threshold h or not. In this case, independently of its current role, v joins u's cluster. Otherwise, if v is a clusterhead with more than k neighbors which are clusterheads, as in the case of a new link, the weight of the clusterhead x that violates the k-neighborhood condition is determined, and correspondingly the clusterhead with the smallest weight will resign.

On receiving CH(u);
begin
    if $(w_u > w_{Clusterhead} + h)$ **then begin**
                        send JOIN(v,u);
                        $Clusterhead := u;$
                        if $Ch(v)$ **then** $Ch(v) := $ **false**
                end
                else if $Ch(v)$ **and** $|\{z \in \Gamma(v) : Ch(z)\}| > k$
                    then begin
                        $w := \min_k\{w_z : z \in \Gamma(v) \wedge Ch(z)\};$
                        if $w_v > w$ **then send** RESIGN(w)
                        else begin
                            $x := \max_{w_z > w_v}\{z : Ch(z)\};$
                            send JOIN(v,x);
                            $Clusterhead := x;$
                            $Ch(v) := $ **false**
                        end
                end

  end;

- *On receiving* JOIN(u,z). On receiving the message JOIN(u,z), the behavior of node v depends on whether it is a clusterhead or not. In the affirmative, v has to check if either u is joining its cluster ($z = v$: in this case, u is added to $Cluster(v)$) or if u belonged to its cluster and is now joining another cluster ($z \neq v$: in this case, u is removed from $Cluster(v)$). If v is not a clusterhead, it has to check if u was its clusterhead. Only if this is the case, v has to decide its role: It will join the biggest clusterhead x in its neighborhood such that $w_x > w_v$ if such a node exists. Otherwise, it will be a clusterhead. In this latter case, if the k-neighborhood condition is violated, a RESIGN message is transmitted in order for the clusterhead with the smallest weight in v's neighborhood to resign.

On receiving JOIN(u, z);
begin
   if $Ch(v)$
     then if $z = v$ **then** $Cluster(v) := Cluster(v) \cup \{u\}$
          else if $u \in Cluster(v)$ **then** $Cluster(v) := Cluster(v) \setminus \{u\}$
   else if $Clusterhead = u$ **then**
        if $\{z \in \Gamma(v) : w_z > w_v \wedge Ch(z)\} \neq \emptyset$
        then begin
             $x := \max_{w_z > w_v}\{z : Ch(z)\}$;
             send JOIN(v,x);
             $Clusterhead := x$
        end
      else begin
           send CH(v);
           $Ch(v) := $ **true**;
           $Clusterhead := v$;
           $Cluster(v) := \{v\}$;
           if $|\{z \in \Gamma(v) : Ch(z)\}| > k$ **then**
              send RESIGN$(\min_k\{w_z : z \in \Gamma(v) \wedge Ch(z)\})$
      end
end;

• *On receiving* RESIGN(w). On receiving the message RESIGN(w), node v checks if its weight is $\leq w$. In this case, it has to resign and it will join the neighboring clusterhead with the biggest weight. Notice that since the M-procedures are supposed to be not interruptible, and since v could have resigned already, it has also to check if it is still a clusterhead.

On receiving RESIGN(w);
begin
   if $Ch(v)$ **and** $w_v \leq w$ **then begin**
                $x := \max_{w_z > w_v}\{z : Ch(z)\}$;
                send JOIN(v,x);
                $Clusterhead := x$;
                $Ch(v) := $ **false**
        end
end;

The correctness of the described protocol in achieving the ad hoc clustering properties listed above can be found in [21] along with simulation results that demonstrate the effectiveness of the protocol in reducing the overhead of role switching in presence of the mobility of the nodes.

4 Forming Ad Hoc Networks of Bluetooth Devices

In this section we illustrate the use of clustering as described in the previous section to define a protocol for *scatternet formation*, i.e., the formation of an ad hoc network of Bluetooth devices. The protocol outlined in this section is joint research with Professor Chiara Petrioli and has been described more thoroughly in [24] and [25].

Bluetooth Technology (BT) [26] is emerging as one of the most promising enabling technologies for ad hoc networks. It operates in the 2.4GHz, unlicensed ISM band, and adopts frequency hopping spread spectrum to reduce interferences.

When two BT nodes come into each others communication range, in order to set up a communication link, one of them assumes the role of *master* of the communication and the other becomes its *slave*. This simple "one hop" network is called a *piconet*, referred in the following as a *BlueStar*, and may include many slaves, no more than 7 of which can be active (i.e., actively communicating with the master) at the same time. If a master has more than seven slaves, some slaves have to be "parked." To communicate with a parked slave a master has to "unpark" it, while possibly parking another slave.

All active devices in a piconet share the same channel (i.e., a frequency hopping sequence) which is derived from the unique ID and Bluetooth clock of the master. Communication to and from a slave device is always performed through its master.

A BT device can timeshare among different piconets. In particular, a device can be the master of one piconet and a slave in other piconets, or it can be a slave in multiple piconets. Devices with multiple roles will act as gateways to adjacent piconets, resulting in a multihop ad hoc network called a *scatternet*.

Although describing methods for device discovery and for the participation of a node to multiple piconets, the BT specification does not indicate any methods for scatternet formation. The solutions proposed in the literature so far ([27], [28], and [29]), either assume the radio vicinity of all devices ([27] and [29]), or require a designated device to start the scatternet formation process, [28]. Furthermore, the resulting scatternet topology is a tree, which limits the efficiency and robustness of the resulting scatternet.

In this paper we present BlueStars, a new scatternet formation protocol for multi-hop Bluetooth networks, that overcomes the drawbacks of previous solutions in that it is fully distributed, does not require each node to be in the transmission range of each other node and generates a scatternet whose topology is a mesh rather than a tree.

The protocol proceeds in three phases:

1. The first phase, *topology discovery*, concerns the *discovery of neighboring devices*. By the end of this phase, neighboring devices acquire a "symmetric" knowledge of each other.
2. The second phase takes care of BlueStar (piconet) formation. By the end of this phase, the whole network is covered by disjoint piconets.
3. The final phase concerns the selection of *gateway devices* to connect multiple BlueStars so that the resulting *BlueConstellation* is connected.

These three phases are described in the following sections.

4.1 Topology Discovery

The first phase of the protocol, the topology discovery phase, allows each device to become aware of its one hop neighbors' ID and weight. According to the BT specification version 1.1, discovery of unknown devices is performed by means of the *inquiry procedures*.

The problem of one-hop neighbors discovery in Bluetooth has been dealt with extensively in [27] (for "single hop" networks, i.e., networks in which all devices are in each other transmission range) and [30] (for multihop networks). The BT inquiry and paging procedures are used to set up two-node temporary piconets through which two neighboring devices exchange identity, weight and synchronization information needed in the following phases of the scatternet formation protocol. This information exchange allows a "symmetric" knowledge of one node's neighbors, in the sense that if a node u discovers a neighbor v, node v discovers u as well.

4.2 BlueStars Formation

In this section, we describe a distributed protocol for grouping the BT devices into piconets. Given that each piconet is formed by one master and a limited number of slaves that form a star-like topology, we call this phase of the protocol BlueStars formation phase.

Based on the information gathered in the previous phase, namely, the ID, the weight, and synchronization information of the discovered neighbors, each device performs the protocol locally. The rule followed by each device is the following: A device v decides whether it is going to be a master or a slave depending on the decision made by the neighbors with bigger weight (v's "bigger neighbors"). In particular, v becomes the slave of the first master among its bigger neighbors that has paged it and invited it to join its piconet. In case no bigger neighbors invited v, v itself becomes a master. Once a device has decided its role, it communicates it to all its (smaller) neighbors so that they can also make their own decision.

Let us call *init devices* all the devices that have the biggest weight in their neighborhood. If two nodes have the same weight, the tie can be broken by using the devices unique ID. Init devices are the devices that initiate the BlueStars formation phase. They will be masters. As soon as the topology discovery phase is over, they go to page mode and start paging their smaller neighbors one by one. All the other devices go in paging scan mode.

The protocol operations in this phase are described by the initOperations() procedure described below.

```
initOperations() {
    if (for each neighbor u: myWeight > uWeight) {
        myRole = 'master';
        go to page mode;
        send page(v, master, v) to all smaller neighbors;
        exit the execution of this phase of the protocol; }
```

```
    else
        go to page scan mode;
}
```

The following procedure is triggered at a non-init device v by the reception of a page. The parameter of the page are the identity of the paging device u, its role (either 'master' or 'slave') and, in the case the paging device u is a slave, the identity of the device to which it is affiliating. (In case u is a master this parameter is irrelevant and can be set to u itself.)

```
onReceivingPage(deviceId u, string role, deviceId t) {
record that u has paged;
record role(u);
if (role(u) == 'slave')
    master(u) = t;
if (myWeight < uWeight) {
    if (role(u) == 'master')
        if (myRole == 'none') {
            join u's piconet;
            myMaster = u;
            myRole = 'slave'; }
        else
            inform u about myMaster's ID;
    if (some bigger neighbor has to page yet)
        exit and wait for the following page;
    else {
        switch to page mode;
        if (all bigger devices are slaves) {
            myRole = 'master';
            send page(v, master, v) to each neighbors
            (smaller neighbors first);
            exit the execution of this phase of the protocol; }
        else {
            send page(v, slave, myMaster) to each neighbors;
            switch to page scan mode; } } }
else
    if (all neighbors have paged)
        exit the execution of this phase of the protocol;
    else
        exit and wait for the next page;
}
```

The procedure of recording the role of a device u includes all the information of synchronization, addressing, etc., that enable v to establish a communication with u at a later time, if needed.

Upon receiving a page from a device u, device v starts checking if this is a page from a bigger neighbor or from a smaller one. In the former case, it checks if the sender of the page is a master. If so, and v is not part of any piconet yet, it joins device u's piconet. If instead device v has already joined a piconet, it informs device u about this, also communicating the name of its master. Device v then proceeds to check if all its bigger neighbors have paged it. If this is not the case, it keeps waiting for another page (exiting the execution of the procedure).

When successfully paged by all its bigger neighbors, device v knows whether it has already joined the piconet of a bigger master or not. In the first case, device v is the slave of the bigger master that paged it first. In the latter case device v itself is going to be a master. In any case, device v goes to page mode, and communicates its decision to all its smaller neighbors.

At this point, a master v exits the execution of this phase of the protocol. If device v is a slave, it returns to page scan mode and waits for pages from all its smaller neighbors of which it still does not know the role. Indeed, some of a slave's smaller neighbors may not have decided their role at the time they are paged by the bigger slave. As soon as a device makes a decision on its role, it therefore pages its bigger slaves and communicates whether it is a master or a slave, along with its master ID (if it is a slave). This exchange of information is necessary to implement the following phase of gateway selection for obtaining a connected scatternet (see Section 4.3).

Notice that the outermost else is executed only by a slave node, since once it has paged all its neighbors, a master has a complete knowledge of its neighbors role and of the ID of their master and thus it can quit the execution of this phase of the protocol.

Implementation in the Bluetooth Technology. The protocol operations of this phase all rely on the standard Bluetooth paging procedures. However, the paging and paging scan procedure described above assume the possibility of exchanging additional information, namely, a device role and for slaves, the ID of their masters. These information cannot be included in the FHS packet which is the packet exchanged in the standard paging procedures.

Our proposal is to add an LMP protocol data unit (PDU), including fields to record the role of the sending device and the ID of its master, to easily exchange the information needed for scatternet formation while possibly avoiding a complete set up of the piconet.

Of course, whenever a slave joins a non-temporary piconet, a complete piconet set up has to be performed, after which the slave is put in park mode to allow it to proceed with the protocol operation (e.g., performing paging itself).

4.3   Configuring BlueStars

The purpose of the third phase of our protocol is to interconnect neighboring BlueStars by selecting inter-piconet gateway devices so that the resulting scatternet, a *BlueConstellation*, is connected whenever physically possible. The main task accomplished by this phase of the protocol is gateway selection and interconnection.

Two masters are said to be *neighboring masters* (*mNeighbors*, for short) if they are at most three hops away, i.e., if the shortest path between them is either a two-hops path (there is only one slave between the two masters) or a three-hops path (there are two slaves).

A master is said to be an *init master*, or simply an *iMaster*, if it has the biggest weight among all its mNeighbors. Therefore, the set of masters that results from the BlueStars formation phase is partitioned into two sets, the iMasters and the non-iMasters devices.

The connectivity of the scatternet is guaranteed by a result, first proven in [31], that states that given the piconets resulting from the BlueStars formation phase, a BlueConstellation—a connected BT scatternet—is guaranteed to arise if each master establishes multihop connections to all its mNeighbors. These connections are all needed to ensure that the resulting scatternet is connected, in the sense that if any of them is missing the scatternet may be not connected.

This result provides us with a criterion for selecting gateways that ensures the connectivity of the resulting scatternet: all and only the slaves in the two and three-hops paths between two masters will be gateways. If there is more than one gateway device between the same two masters they might decide to keep only one gateway between them, or to maintain multiple gateways between them.

Upon completion of the previous phase of the protocol a master v is aware of all its mNeighbors. It directly knows all its neighboring slaves which in turn are aware of (and can communicate to the master v) the ID of their master and of the master of their one-hop slave neighbors.

Establishment of a connected scatternet. We are finally able to establish all the connections and the needed new piconets for obtaining a BlueConstellation, i.e., a connected scatternet.

This phase is initiated by all masters v by executing the following procedure.

```
mInitOperations() {
if (for each mNeighbor u: myWeight > uWeight) {
     myRole = 'iMaster';
     instruct all gateway slaves about which neighbors to page;
     go to page mode;
     page all the slaves which belong to a different piconet
          and have been selected as interconnecting devices;
     exit the execution of this phase of the protocol; }
else {
     tell all gateway slaves to bigger mNeighbors
          to go to paging scan mode;
     if (there are bigger mNeighbors' slaves in my neighborhood
          which will interconnect the two piconets)
               go to page scan mode;
     tell all gateway to smaller mNeighbors to go to paging mode
```

```
        when the links to bigger mNeighbors are established;
  if (there are smaller mNeighbors' slaves in my neighborhood
        which will interconnect the two piconets)
        go to page mode when the links to bigger mNeighbors are up; }
}
```

Every master v starts by checking whether it is an iMaster or not. If it is an iMaster, then it instructs each of its gateway slaves to go into page mode and to page (if any):

- Its two-hop mNeighbors. In this case, as soon as v's slave has become the master of an mNeighbor u, they perform a switch of roles, as described in the BT specification, so that v's slave become also a slave in u's piconet. In this case, no new piconet is formed and the slave in between u and v is now a slave in both their piconets, as desirable.

- The slaves of its three-hop mNeighbors (that are two-hops away from v). In this case v's slave becomes also a master of a piconet whose slaves are also slaves to the three-hop mNeighbors, i.e., a new piconet is created to be the *trait d'union* between the two masters.

The iMaster v itself can then go into paging mode to recruit into its piconet some of those neighboring slaves (if any) that joined some other piconets, so that these slaves can be the gateway to their original masters.

Notice that, given the knowledge that every master has about its "mNeighborhood," an iMaster v instructs each of its gateway slaves about exactly who to page, and the resulting new piconet composition. If, for instance, a slave is gateway to multiple piconets, iMaster v knows exactly to which of the neighboring piconet its slave is going to be also a slave, and if it has to be master of a piconet that can have, in turn, multiple slaves.

When the gateway slaves of a non-iMaster device v have set up proper connections toward bigger mNeighbors, they will go into page mode and page those of its two-hop mNeighbors and of the the slaves of its three-hop mNeighbors with which they have been requested by v to establish a connection.

Implementation in the Bluetooth Technology. The mechanism described above can be easily implemented by means of the BT standard procedures for parking and unparking devices, and those for link establishment. In particular, upon completion of the second phase of the protocol, a slave asks its master to be unparked. The master will then proceed activating (unparking) different groups of slaves, and collecting from them all the information required for configuring the BlueConstellation. Based on this information, the master will then make a decision on which links to establish to connect with its mNeighbors, and will unpark the gateways in groups of seven to inform them of the piconets to which they are gateway. Each gateway will then run the distributed procedure for interconnecting neighboring piconets described in the previous section, at the end of which it will issue to the master a request for being unparked in order to communicate the list of links successfully established.

5 Conclusions

In this paper we have described some issues and solutions proposed for ad hoc networking. In particularly, we have illustrated leading MAC protocols, clustering protocols and we have shown how these protocols can be applied to networks of Bluetooth devices for the formation of Bluetooth scatternet.

References

1. E. M. Royer and C.-K. Toh. A review of current routing protocols for ad hoc mobile wireless networks. *IEEE Personal Communications*, 6(2):46–55, April 1999.
2. S. Giordano, I. Stojmenovic, and L. Blazevic. Position based routing algorithms for ad hoc networks: A taxonomy. http://www.site.uottawa.ca/ ivan/wireless.html, 2002.
3. A. D. Myers and S. Basagni. *Wireless Media Access Control*, chapter 6, pages 119–143. John Wiley and Sons, Inc., 2002. I. Stojmenovic, Ed.
4. F. A. Tobagi and L. Kleinrock. Packet switching in radio channels. ii. the hidden terminal problem in carrier sense multiple-access and the busy-tone solution. *IEEE Transactions on Communications*, COM-23(12):1417–1433, December 1975.
5. C. Wu and V. O. K. Li. Receiver-initiated busy tone multiple access in packet radio networks. *ACM Computer Communication Review*, 17(5):336–342, August 1987.
6. A. Gummalla and J. Limb. Wireless collision detect (WCD): Multiple access with receiver initiated feedback and carrier detect signal. In *Proceedings of the IEEE ICC 2000*, volume 1, pages 397–401, New Orleans, LA, June 2000.
7. P. Karn. Maca—a new channel access protocol for packet radio. In *Proceedings of ARRL/CRRL Amateur Radio 9th Computer Networking Conference*, pages 134–140, 1990.
8. V. Bharghavan and al. MACAW: A media access protocol for wireless LANs. *ACM Computer Communication Review*, 24(4):212–225, October 1994.
9. C. Lin and M. Gerla. Real-time support in multihop wireless networks. *ACM/Baltzer Wireless Networks*, 5(2):125–135, 1999.
10. F. Talucci and M. Gerla. MACA-BI (MACA by invitation): A wireless MAC protocol for high speed ad hoc networking. In *Proceedings of IEEE ICUPC'97*, volume 2, pages 913–917, San Diego, CA, October 1997.
11. I. Chlamtac and A. Faragó. Making transmission schedule immune to topology changes in multi-hop packet radio networks. *IEEE/ACM Transactions on Networking*, 2(1):23–29, February 1994.
12. R. Lidl. *Introduction to Finite Fields and Their Applications*. Cambridge University Press, 1994.
13. S. Basagni and D. Bruschi. A logarithmic lower bound for time-spread multiple-access (TSMA) protocols. *ACM/Kluwer Wireless Networks*, 6(2):161–163, May 2000.
14. S. Basagni, D. Bruschi, and I. Chlamtac. A mobility transparent deterministic broadcast mechanism for ad hoc networks. *ACM/IEEE Transactions on Networking*, 7(6):799–807, December 1999.
15. S. Basagni, A. D. Myers, and V. R. Syrotiuk. Mobility-independent flooding for real-time, multimedia applications in ad hoc networks. In *Proceedings of 1999 IEEE Emerging Technologies Symposium on Wireless Communications & Systems*, Richardson, TX, April 12–13 1999.

16. D. Bruschi and M. Del Pinto. Lower bounds for the broadcast problem in mobile radio networks. *Distributed Computing*, 10(3):129–135, April 1997.

17. I. Cidon and M. Sidi. Distributed assignment algorithms for multihop packet radio networks. *IEEE Transactions on Computers*, 38(10):1353–1361, October 1989.

18. C. Zhu and S. M. Corson. A five-phase reservation protocol (FPRP) for mobile ad hoc networks. In *Proceedings of IEEE Infocom'98*, volume 1, pages 322–331, San Francisco, CA, March/April 1998.

19. A. Ephremides and T. V. Truong. Scheduling broadcasts in multihop radio networks. *IEEE Transactions on Communications*, 38(4):456–460, April 1990.

20. S. Basagni. Distributed clustering for ad hoc networks. In A. Y. Zomaya, D. F. Hsu, O. Ibarra, S. Origuchi, D. Nassimi, and M. Palis, editors, *Proceedings of the 1999 International Symposium on Parallel Architectures, Algorithms, and Networks (I-SPAN'99)*, pages 310–315, Perth/Fremantle, Australia, June 23–25 1999. IEEE Computer Society.

21. S. Basagni. Distributed and mobility-adaptive clustering for multimedia support in multi-hop wireless networks. In *Proceedings of the IEEE 50th International Vehicular Technology Conference, VTC 1999-Fall*, volume 2, pages 889–893, Amsterdam, The Netherlands, September 19–22 1999.

22. M. Chatterjee, S. K. Das, and D. Turgut. An on-demand weighted clustering algorithm (WCA) for ad hoc networks. In *Proceedings of IEEE Globecom 2000*, San Francisco, CA, November 27–December 1 2000. To appear.

23. S. Basagni. A note on causal trees and their applications to CCS. *International Journal of Computer Mathematics*, 71:137–159, April 1999.

24. S. Basagni and C. Petrioli. Scatternet formation protocols. Technical report, Università di Roma "La Sapienza", Roma, Italy, September 2001.

25. S. Basagni and C. Petrioli. A scatternet formation protocol for ad hoc networks of Bluetooth devices. In *Proceedings of the IEEE Semiannual Vehicular Technology Conference, VTC Spring 2002*, Birmingham, AL, May 6–9 2002.

26. http://www.bluetooth.com. *Specification of the Bluetooth System, Volume 1, Core.* Version 1.1, February 22 2001.

27. T. Salonidis, P. Bhagwat, L. Tassiulas, and R. LaMaire. Distributed topology construction of Bluetooth personal area networks. In *Proceedings of the IEEE Infocom 2001*, pages 1577–1586, Anchorage, AK, April 22–26 2001.

28. G. Záruba, S. Basagni, and I. Chlamtac. Bluetrees—Scatternet formation to enable Bluetooth-based personal area networks. In *Proceedings of the IEEE International Conference on Communications, ICC 2001*, Helsinki, Finland, June 11–14 2001.

29. C. Law, A. K. Mehta, and K.-Y. Siu. Performance of a new Bluetooth scatternet formation protocol. In *Proceedings of the 2001 ACM International Symposium on Mobile Ad Hoc Networking & Computing, MobiHoc 2001*, pages 183–192, Long Beach, CA, 4–5 October 2001.

30. S. Basagni, R. Bruno, and C. Petrioli. Device discovery in Bluetooth networks: A scatternet perspective. In *Proceedings of the Second IFIP-TC6 Networking Conference, Networking 2002*, Pisa, Italy, May 19–24 2002.

31. I. Chlamtac and A. Faragó. A new approach to the design and analysis of peer-to-peer mobile networks. *Wireless Networks*, 5(3):149–156, May 1999.

Communications through Virtual Technologies

Fabrizio Davide[1], Pierpaolo Loreti[2], Massimiliano Lunghi[1],
Giuseppe Riva[3], and Francesco Vatalaro[2]

[1] Telecom Italia, Rome, Italy
fabrizio.davide@telecomitalia.it, ue001862@guest.telecomitalia.it
[2] Università di Roma Tor Vergata, Rome, Italy
{loreti, vatalaro}@ing.uniroma2.it
[3] Universitàà Cattolica del Sacro Cuore di Milano, Milan, Italy
giuseppe.riva@mi.unicatt.it

Abstract. The evolution of technology in support to the Knowledge Society of the years 2000s will be driven by three dominant trends: a) increase of richness and completeness of communications towards "Immersive Virtual Telepresence" (IVT), including an increased attention to the aspects of human perception and of person-machine interaction; b) increasingly relevant role of mobility starting from the UMTS "Beyond 3rd Generation" (B3G); c) pervasive diffusion of intelligence in the space around us through the development of network technologies towards the objective of "Ambient Intelligence" (AmI). This tutorial paper intends to provide an insight on some of the main technologies that will allow the development of this new perspective, as well as on perception issues and related human factors. The paper also provides information on ongoing and planned projects on virtual technologies and ambient intelligence technologies.

1 Introduction

The advent of the Knowledge Society of years 2000s and 2010s is destined to change the way people interact between themselves and with objects around them. A new role for the Information Technology (IT) is already visible, while technology's focus is gradually shifting away from the computer as such to the user [1]. This change of paradigm has the objective to make communication and computer systems simple, collaborative and immanent. Interacting with the environment where they work and live, people will naturally and intuitively select and use technology according to their own needs. One first sign of this change has been the creation of totally new interactive communication environments, such as Computer Mediated Communication (CMC) and Computer Supported Collaborative Work (CSCW). Appeared around twenty years ago, these communication environments are now very common in offices, just behind word processing applications [2].

The evolutionary technology scenarios in support of the Knowledge Society of the years 2000s will be driven by three dominant trends:

E. Gregori et al. (Eds.): Networking 2002 Tutorials, LNCS 2497, pp. 124–154, 2002.
© Springer-Verlag Berlin Heidelberg 2002

- increase of richness and completeness of communications, through the development of multimedia technologies, towards "Immersive Virtual Telepresence" (IVT), including an increased attention to the aspects of human perception and of person-machine interaction;
- increasingly relevant role of mobility, through the development of mobile communications, moving from the Universal Mobile Telecommunications System (UMTS) "Beyond 3rd Generation" (B3G);
- pervasive diffusion of intelligence in the space around us, through the development of network technologies towards the objective of the so-called "Ambient Intelligence" (AmI).

The next IT "wave" will determine the convergence of IVT technologies, B3G mobile communications and ambient intelligence technologies (Figure 1). This

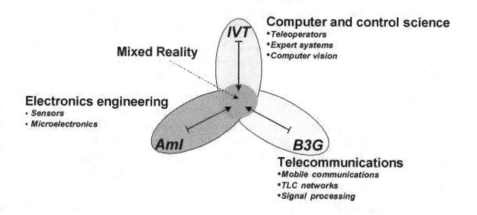

Fig. 1. Converging technologies.

convergence will produce the advent of ubiquitous 3-D telecommunications and the appearance of intelligent environments in which complex multimedia contents integrate and enrich the real space. The most ambitious objective is Mixed Reality (MR) to seamlessly integrate computer interfaces into the real environment, so that the user can interact with other individuals and with the environment itself in the most natural and intuitive way. Mixed reality requires the concurrent development of different disciplines ranging from engineering to ergonomics, from communications to psychology.

In the paper we attempt to outline the link between technology and human factors, to identify the main avenues of the development of virtual technologies for telecommunications, and to report on some of the most important research projects going on and planned.

2   Presence and Telepresence in Virtual Communications

2.1   From Teleoperator to Immersive Telepresence

In the area of CMC and CSCW, a very interesting application is the teleoperator. A teleoperator is a robotic device that is controlled by a human operator through sensors and actuators to function in remote, inaccessible or hazardous, environments. Common to all teleoperation systems is a human operator in a manual or supervisory control loop overseeing performance of remote task. Application areas include operations in radioactive, underwater, space, surgical, disability assistive and "clean room" environments.

In teleoperator systems, the remote manipulator system can be augmented with semiautonomous functions which can either be preprogrammed, taught by demonstration or self-learning. An operator interface at the local site supports human sensory information processing and decision-making. Augmentation at the operator interface could include graphic, multimedia or Virtual Reality (VR) display, text or animation, all presented in an intuitive manner [3].

The development of teleoperators is a key point in the evolution of the person-computer interaction [4]. Teleoperation produces a relationship between the user and the computer, in which technology provides a prosthesis to the user, extending his/her capabilities.

A step forward in the person-computer interaction is telepresence. Telepresence can be described as the ideal of collecting sufficient information through computer means and communicating this to a person in a sufficiently natural way, so that he/she feels to be physically present at the remote site [3].

Even though the most used CMC application is e-mail, technology is allowing the development of more advanced communication environments. Virtual Reality (VR) can be considered as the leading edge of a general evolution of present communication interfaces like telephone, television and computer. The ultimate objective of this evolution is the full *immersion* of the human sensorimotor channels into a vivid and global communication experience [5].

But what is virtual reality? VR is sometimes described as a particular collection of technological hardware. People identify VR with a collection of devices: a computer capable of 3D real-time animation, a head-mounted display, data gloves equipped with one or more position trackers, etc. However, this exclusive focus on technology is certainly disappointing [6,7]. As noted by Steuer this approach "fails to provide any insight into the processes or effects of using these systems, fails to provide a conceptual framework from which to make regulatory decisions and fails to provide an aesthetic from which to create media products" [8].

If VR cannot be simply reduced to a collection of hardware, where should we look to identify its "essence"? According to Bricken [9] the essence of VR is the inclusive relationship between the participant and the virtual environment, where direct experience of the immersive environment constitutes communication.

2.2 Concepts of Presence and Telepresence

Following this approach, VR can be defined in terms of human experience as "a real or simulated environment in which a perceiver experiences telepresence," where telepresence can be described as the "experience of presence in an environment by means of a communication medium" [8].

For researchers working on Virtual Environments (VEs) or Teleoperation Systems (TSs) a clear definition of presence and telepresence may be useful to orient their work. Most researchers in these fields share the current meaning assigned to these concepts, which are defined by Schloerb as follows: physical presence designates "...the existence of an object in some particular region of space and time. For example, this text (in some form) is physically present in front of you now". According to Schloerb, physical presence supports subjective presence, consisting of the perception of being located in the same physical space in which a certain event occurs, a certain process takes place, or a certain person stands [10].

Following this approach, the key difference between presence and telepresence is that the former is a "natural" fact, whereas the latter is a fact produced by technology, an artifact. This difference also occurs in the definition of Slater and Wilbur [11], according to whom presence "...is both a subjective and objective description of a person's state with respect to an environment". Particularly, objective description is defined as "...an observable behavioral phenomenon, the extent to which individuals behave in a VE similar to the way they would behave in analogous circumstances in everyday reality" [11].

Even if these definitions seem very intuitive to us, their effects are not. In fact, there continues to be contention and general disagreement on the role of presence and telepresence in a VE. Specifically, this position has an important implication for the design of a VE: the adequacy of a telepresence system depends on the fidelity in recreating conditions that allow us to perceive ourselves, or other people or objects, as physically present in the "real" environment. Most of the research tries to improve presence of a VE by providing to the user a more "realistic" experience, such as adding physical attributes to virtual objects. For instance, Hoffman et al. published the results of two experiments in which they tried to verify if adding olfactory cues and tactile feedback to a VE may improve the sense of presence [12].

But is it really so important this focus on the physical characteristics of a VE? According to [13], more than on the richness of available images sense of presence depends on the level of interaction/interactivity which actors have both in the real environment and in the simulated environment. When it is used for real world applications, VE is effective when the user is able to navigate, select, pick up, move and manipulate objects much more naturally.

The roots of this new position are in Heidegger's philosophy and in the theory of perception of J. Gibson [14]. According to these authors, the environment does not provide undifferentiated information, ready-made objects equal for everyone. It offers different opportunities according to the actors and their needs.

Affordances are not "things which are outside" simply waiting for someone to come and take a photograph of them. They are resources, which are only revealed to those who seek them. What has all this to do with presence, telepresence and virtual environments? Zahoric and Jenison [14] explain it clearly: "*presence is tantamount to successfully supported action in the environment*" (italics in the original).

These considerations led Mantovani and Riva to propose a cultural concept of presence as a social construction [15]. Lying at the basis of this view are two elements which promise a high sense of presence: a cultural framework and the possibility of negotiation of actions and of their meaning [16]. Tracing a surprising parallel between the task of VR designers and that of phone sex workers - two figures who have the task of making the human body visible by means of extremely narrow channels of communication and who succeed in their task to the extent in which they use powerful, shared, cultural codes - Stone [17] describes the context, as composed mainly of symbolic references which allow actors to orient and coordinate themselves. Within this view, experiencing presence and telepresence does not depend so much on the fidelity of the reproduction of 'physical' aspects of 'external reality', as it depends on the capacity of simulation to produce a context in which social actors may communicate and cooperate [16].

Following this approach a new definition of presence can be proposed that (a) recognizes the mediated character of every possible experience of presence; (b) always conceives experience as immersed in a social context; (c) stresses the component of ambiguity inherent in everyday situations; (d) highlights the function of explanation which culture (artifacts and principles) plays. Breaking down this idea into simple formulations, we may say that [16]:

-- presence is always mediated by both physical and conceptual tools which belong to a given culture: "physical" presence in an environment is no more "real" or more true than telepresence or immersion in a simulated virtual environment;
-- the criterion of the validity of presence does not consist of simply reproducing the conditions of physical presence but in constructing environments in which actors may function in an ecologically valid way: we accept the emphasis of ecological approach or the primacy of action on mere perception;
-- action is not undertaken by isolated individuals but by members of a community who face ambiguous situations in a relatively coordinated way: to be able to speak of an actor's presence in a given situation, freedom of movement must be guaranteed, both in the physical environment (locomotion) and in the social environment, composed of other actors involved in the same situation, in whatever way and for whatever reason.

The main consequence of this approach for the design and the development of VR systems is that the users' presence in an environment exists if and only if they can use VR to interact. To allow interaction in a given situation, the user's freedom must be guaranteed, both in the physical and in the social environment.

2.3   Virtual Immersive Cooperative Environments

The first step to achieving mixed reality is the implementation of genuinely effective, Virtual Immersive Cooperative Environments (VICEs). The main feature which distinguishes a VICE from a normal Computer Supported Cooperative Work (CSCW) is the use of a *Shared Space* which represents both the users of the environment and the data they are manipulating [18].

In this setting the metaphor of "the computer as a window" though which users can see objects and other users gives way to the concept of virtual reality as a "flexible shared space" [19]. VICEs offer their users a digital environment which can be populated with users replicas and objects. Users cybernetic replicas are called avatars. In this form they are free to navigate, meet other users and manipulate data usually represented in the form of three dimensional objects. At the same time they can use audio and video channels to interact with other users using both verbal and non-verbal forms of communication.

The designer of a VICE aims at building tools that make it possible for users - individually or in small groups - to work autonomously. At the same time the tools should monitor the effectiveness of the interaction. This implies that the virtual environment has to allow changes in the way in which the user is represented, allowing monitoring what is going on in the environment and moving rapidly from place to place [20]. If I am working on my own I do not want to be disturbed. The other users should know this - otherwise they might send me messages and receive no reply. I want to be able to work in those areas of the environment including the information which interests me. Churchill and Snowdon [18] point out that "the majority of current systems are rather clumsy in monitoring the transition from one form of interaction or one mode of use to another; navigation systems do not allow users to move in the same way they would move in the 'real world'".

Let us look at one example. In early VICEs the main system for navigation and interaction was a glove. If the user pointed in a particular direction the avatar would move in that direction. By clenching the fist or moving fingers together the user could grasp or modify an object. In theory this was very simple. Users' actual experience showed, however, that this was not always so [21]. As a start it was necessary to go through a complex calibration procedure every time the system was used by a new user: adapting the sensitivity of the tool to the needs of the individual user is not easy. Many first time users, for instance, prefer to move slowly while they are getting used to the new experience; an expert user, on the other hand, often prefers to move at maximum speed, reaching the needed information in as little time as possible.

The search for solutions to these problems is a very active field of research. Researchers at the University of Caen have recently produced a conversational agent that allows a user to navigate a VICE using voice commands [22], [23]. This system, called Ulysses, has achieved good results in experimental tests. In particular, the syntactic and semantic analysis system managed to understand complex orders such as: "get into the car in front of the house" or "walk past the flag and go into the house". This allowed users not only to move but to engage in simple forms of interaction with the objects present in the virtual environment.

If the internal organization of the environment, as perceived by the user, depends on a shared interpretation of its "meaning", the designer has to bear in mind that cooperation is always "situated" and that it is continually influenced by specific circumstances and by the interests of specific actors. This has five key ergonomic consequences:

- *Any model of a VICE has to allow actors to develop shared interpretations of its meaning [24].*
  Decision making depends on a negotiated reduction in the gap between different actors' individual frames of reference.
- *Any model of cooperation within a VICE is tightly dependent on the specific application area for which the system has been developed [25].*
  The designer has to create a model in such a way as to reflect the characteristics of the situation and the goals of the actors.
- *Any model of cooperation will influence the way in which users interact with the system [26].*
  When VICE systems introduce new cooperative technologies they inevitably modify interactive processes. As we have already seen on several occasions new technologies are never transparent to users. The designer of a VICE has to realize that the first thing users will do is to try to give a sense to the artifacts they find there. In general terms the users of an VICE face two problems: on the one hand they have to perform their work within the group; on the other hand they have to grip the new technology and learn how to use it appropriately.
- *Any model of cooperation has to make users aware of situations where the model is no longer adequate to lead the interaction [27].*
  In many VICEs individual users receive only a limited degree of feedback. This implies that they are not always capable of evaluating their own performance and the way in which this performance contributes to their general goals. For this reason it is essential that the cooperation model informs users about the way that the system is responding adequately to the situation.
- *Any model of cooperation has to be able to predict the phases in which individual users will organize their work [28].*
  A number of studies have shown how collaborative work in daily life involves alternating phases of group work, individual activity and work in restricted sub-groups. Given that users do not always share the same activities, skills and interests, VICE designers have to provide users with tools to approach group activities in a modular fashion. It is particularly important to create tools to allow individual users or small groups of users to carry out their activities autonomously.

In the Appendix we collect some of the main research activities in Europe in the area of VICE developments.

3 Mobile Radio Technologies

3.1 Evolution Path within 3G

To better understand the basic features expected from $B3G$ systems, and their relationships with virtual technologies and ambient intelligence technologies, we initially depict the main characteristics of $3G$ evolution. Table 1 collects the main characteristics of mobile systems generations, as they are known till $3G$.

Although there is a tendency to refer to the future "beyond $3G$" ($B3G$) as $4G$ or even $5G$, there is a significant discontinuity in aims and modalities in the expected development of technology. In fact, $1G$, $2G$ and $3G$ have been based on the approach of defining and standardizing "monolithic" systems having a certain degree of internal rigidity. On the contrary, $B3G$ is being conceived to be characterized by flexible, reconfigurable and scalable system elements and by the seamless integration of different systems, both mobile and fixed systems.

Flexibility in network interconnection and evolution in network system features will have a limited extent within $3G$. Starting from $3GPP$ ($3G$ Partnership Program) Rel.99 issuing of specifications, a possible evolution path will include at least two additional steps (Figure 2) [29]:

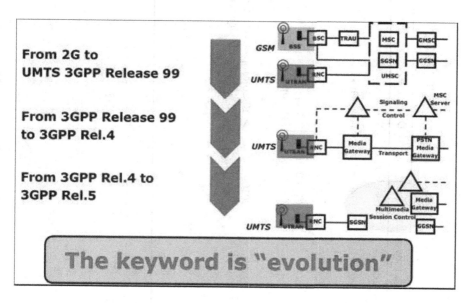

Fig. 2. Evolution path within the UMTS.

– $3GPP$ Rel.04, still with separate transport and signaling control paths, but with unified "Media Gateways" to jointly process circuit switched and packet switched traffic;

Table 1. Wireless generations.

Generation	Acronym	Years	Data rate	Main characteristics
First	2G	1980-1990	N.A.	- analogue voice
Second	2G	1990-2000	Few ten kbps	- digital voice - low rate data - limited roaming capability between networks
Third	3G	2000-2010	2 Mbps (fixed terminals) 144 kbps (vehicular terminals)	- multimedia services - Internet access for mobile terminals - worldwide roaming between mobile networks
Beyond	B3G	2010-	$\geq$ 10-20 Mbps (fixed terminals) $\geq$ 2 Mbps (vehicular terminals)	- integration of systems based on different technologies (not only cellular) - high mobility (radio, network) - seamless coverage - scalability, reconfigurability - terminals involving functionalities for natural interaction

- 3GPP Rel.05, with full network integration of transport and control signals, merging into a single IP paradigm.

This evolutionary approach is suggested by a combination of business and technical reasons.

One of the main concepts in UMTS is that the network establishing connections is separated as much as possible from the network parts creating and maintaining services. To implement this concept $3GPP$ Rel.99 introduces network components, known as service capabilities. In $3GPP$ Rel.99 service capabilities are still proprietary and can be vendor-dependent, and there is no common service creation environment. This is overcome with $3GPP$ Rel.04 and Rel.05 with the Open Service Architecture (OSA). Figure 3 collects some ser-

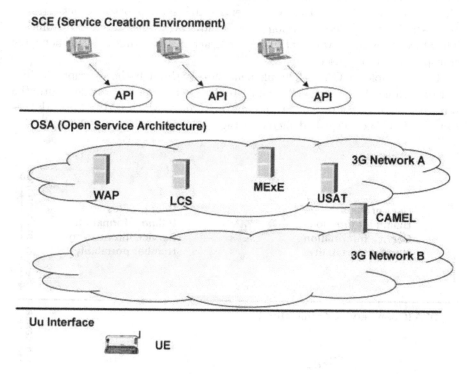

Fig. 3. 3GPP Open Service Architecture.

vice capabilities already defined in $3GPP$ Rel.99 (they will be expanded within Rel.04/05):

- WAP (Wireless Application Protocol), to offer to the end user a web browser tailored to the wireless environment;
- LCS (Location Communication Services), to provide other service capabilities with the user position information;

– MExE (Mobile station application Execution Environment), to provide other service capabilities with information about the ability of the terminal to handle information;
– USAT (UMTS SIM Application Toolkit), to offer the tools required for SIM card handling.

A further service capability of special importance is CAMEL (Customized Applications for Mobile network Enhanced Logic). Within 3G the concept of Intelligent Network (IN) will evolve. The concept of IN is derived directly from the PSTN/ISDN network environment, and thus has some deficiencies when adapted to a mobile network environment. Originally, IN allows transfer of user service information within the user's own home network, but this is not adequate when he/she can roam between different networks. CAMEL is therefore the service capability necessary to transfer service information between networks: it acts as a "service interconnection point" and serves other service capabilities. The objective is to allow completion of transactions involving different networks transparently to the user.

One example of CAMEL implementation is through the concept of VHE (Virtual Home Environment): thanks to VHE the user is provided with the same personalized set of services, no matter through which UMTS network gets connected. This concept is depicted in Figure 4.

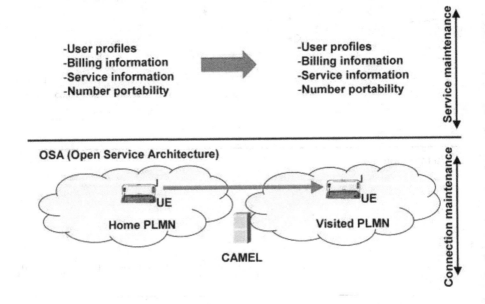

Fig. 4. 3GPP concept of Virtual Home Environment.

3.2 Merging Mobile Communications and Virtual Technologies

According to the Book of Visions 2001 elaborated by the WWRF (Wireless World Research Forum) "*During its first century of existence, the development of communication systems has experienced a serious degradation of end user's ease of use.*" [30]. In fact, the increasing complexity of technology heavily impacts on user friendliness of telecommunications networks.

Conversely, the fast penetration of wireless communications has put into evidence the user's need to get easily connected anywhere and anytime at an affordable price. On the one hand, wireless communications clearly proved that the most a technology provides simple access means, added to freedom of movement and increased security, the most the user is willing to accept it. On the other hand, the most a technology is complex and costly, the less the user is prone to accept it, in spite of possibly large potential advantages, which are generally not reachable by the average user not interested in spending time and energies in acquiring the underlying technology fundamentals. As a consequence, the successful systems of the future will adhere to the paradigm of "disappearing technologies", both valid for communications and computing, and will provide improved ease-of-use at the expense of an increased, but invisible to the user, complexity of the underlying systems and networks necessary to transport and process the information in the different multimedia forms and usage contexts.

As the user must be firmly put at the "center of the universe of technologies", it is clear that the elaboration of a purely technical vision of the future of telecommunications and computing is not only insufficient but even dangerous. Rather, any viable technical solutions must be put into a much wider perspective. According to [30] we mainly need:

- a solid user-centered approach, looking carefully at the new ways users will interact with the wireless systems,
- innovative services and applications made possible with the new technologies,
- new business models and value chain structures overcoming the traditional "user - network provider - service provider" chain.

It is expected that the major innovative thrust will come from new ways of interaction with the wireless system and among systems. The emerging need is bridging the user's real world and virtual world and to continuously put them in contact. The challenge for technology is therefore seamless communication amongst individuals, things, and *cymans* (our synthetic counterparts in the cyberworld - a sort of autonomous avatars) to provide the citizen of the future with a natural enhanced living environment. It is generally considered feasible that this vision may enter into effect around 2010 [30,31].

In the novel user-centric approach being developed for *B3G* there are clearly four main issues to be covered:

- integration of mobile technologies,
- user modeling,
- interaction applications,
- interaction interfaces.

Integration of mobile technologies. It is useful to rely upon a reference model for the interaction between the user and surrounding technologies. In Fig-

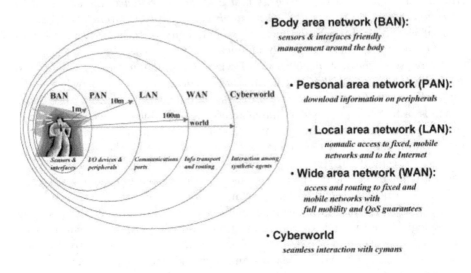

• **Body area network (BAN):**
sensors & interfaces friendly management around the body

• **Personal area network (PAN):**
download information on peripherals

• **Local area network (LAN):**
nomadic access to fixed, mobile networks and to the Internet

• **Wide area network (WAN):**
access and routing to fixed and mobile networks with full mobility and QoS guarantees

• **Cyberworld**
seamless interaction with cymans

Fig. 5. The individual space layers.

ure 5 we show a conceptual layering of the individual space, where we recognize the following main elements:

- Body area network (BAN), for the friendly management of sensors and interfaces in contact with the body and around it;
- Personal area network (PAN), for downloading of information on personal and shared peripherals;
- Local area network (LAN), for the nomadic access to fixed and mobile networks, and to the Internet;
- Wide area network (WAN), for the access and routing with full mobility and the desired degree of QoS (includes mobile and fixed networks);
- "The Cyberworld", where users avatars and cymans seamlessly interact.

All the layers must be integrated to cooperate in service provision to the user. The outmost layer represents the cyberworld, where avatars of the users can stay in touch with other synthetic agents, so to get services and perform transactions. Wireless technologies are the key elements for allowing this seamless interaction inside the real world and between it and the cyberworld. WWRF is elaborating the so-called MultiSphere Reference Model, with similar structure as the model referred above [30].

User modeling. Most of the system and application design today is technology-driven only because we do not have yet the tools to incorporate user behavior as a parameter in system design and product development. A strong effort must be made in the direction of user modeling to achieve in user understanding the same level of confidence we have reached in modeling technology.

The main workhorse has been identified in the development of user scenarios. This is an activity very marginally involving engineering skills: rather, the focal competencies can be provided by a wide set of professional categories (including psychologists, movie directors, science fiction writers, etc.). The logical flow for assessment of viable scenarios is depicted in Figure 6.

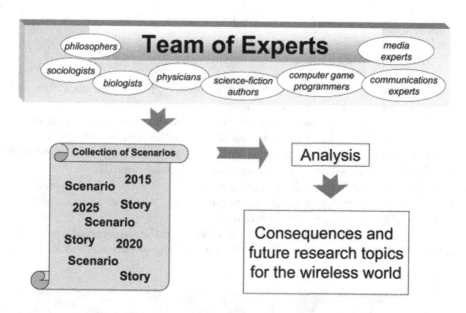

Fig. 6. Logical flow for scenarios modeling. (From [30])

Scenarios must be designed to encompass societal, economic as well as technology developments and form a logical framework in which use cases can be fitted. The European Commission and research organizations, such as the WWRF, encourage scenario-based approaches for pushing the research in the right way [30]. Experts have to analyze the scenarios drawing consequences and future research topics. The main output of these modeling efforts will consist in the "pieces of technology" needed to provide the functionalities envisaged within the reference scenarios.

One main result of user modeling efforts is the user-centric context definition. Much more than simple "user profiling" and "service profiling", abstract user context definition will provide the description of different observable dimensions (i.e. attributes) characterizing a situation in which the user is immersed.

The context oriented systems will be able to answer questions, such as: Where are you? Whom are you with? What resources are close to you? What is your mood today? Defining abstract context types, the "context metadata" (such as home, office, travel, distress, etc.), will help assisting the user in different environments. Dynamical management of context metadata will be based on creation of a database of common abstract context types. Needed context information will be also a function of local culture and on society organization.

Interaction applications. According to [32] the following are some human-cyberworld interaction applications:

- Smart signs added to the real world;
- Information assistant (or "virtual guide");
- AR combined with conversational multimedia.

Smart signs overlaid on user real world may provide information assistance and advertisement based on user preferences.

The virtual guide knows where the user is, his/her heading, as well as the properties of the surrounding environment; interaction can be through voice or gestures, and the virtual guide can be an animated guide and provide assistance in different scenarios based on location and context information.

Conversational multimedia can be also added to a VR scenario, as shown in Figure 7, where a user can see the avatar of another user coming into the scene and a $3 - D$ video conference is carried on.

In the future the terminal will be able to sense the presence of a user and calculate his/her current situation. Throughout the environment, bio-sensing

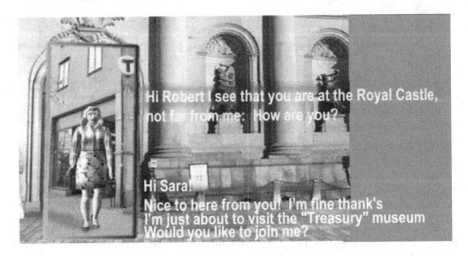

Fig. 7. A scenario of augmented reality combined with $3 - D$ video conference (from [32])

will be used not only for entertainment and medical reasons, but also to enhance person-to-person and person-to-device communications. Biometrics technology will be used to enhance security by combining static (facial recognition) and dynamic information (voice and lip movement, uncontrolled user gestures), as well as user's habits, which the network will be able to acquire and maintain.

Interaction interfaces. Recent advances in wearable computing and intelligent fabrics promise to realize truly ubiquitous communication and computing.

While technology used in everyday life is increasingly complex, the basic human capabilities evolve very slowly. In developing new interaction technologies limitations in sensory and motion ability, in short and long term memory, as well as in brain processing power must be taken into account. In order to fully exploit the technology offer we need to extend the interaction with more senses (touch, smell, and taste) and at the same time make better use of the senses used today (hearing and vision) by also exploring peripheral vision and ambient listening. With even a longer perspective we may consider combining perception through multiple senses and "amplifying" them with new technology protheses.

In the short term the increase of efficient use in human interaction interfaces will be the first step, essentially extending to communication and computing the most intuitive and well assessed means of use of senses and of elementary devices of common use. In the following we present two examples of new interaction interfaces. The first is the phone glove of the Ericsson Research Usability and Interaction Lab, which provides an intuitive extension of the touch sense to communications (Figure 8) [30,33].

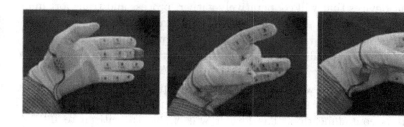

Fig. 8. The phone glove: intuitive use of a sense. (From [30])

The second example refers to extension of functionalities through objects of common use. The pen in Figure 9 at first glance is a conventional pen to write on a piece of paper.

However, it contains a video camera that picks up and transmits the written text. It also contains a set of sensors able to feel how much the nib is being pressed, and the way the pen is handled. If used on a special paper containing invisible identification it is able to associate what is written with the location of writing [34,35].

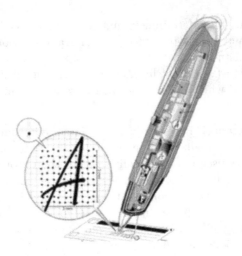

Fig. 9. Chatpen^{TM}, the transmitting pen: extended use of an elementary object. (From [34])

4 Ambient Intelligence

4.1 Ambient Intelligence Scenarios

Ambient Intelligence (AmI), as defined by the Information Society Technologies Advisory Group (ISTAG) to the European Commission, provides a vision of the Knowledge Society [36] of 2010s. In this vision all the environment around us, homes and offices, cars and cities, will collectively realize a pervasive network of intelligent devices that will cooperatively gather, process and transport information. Therefore, in this sense AmI is the direct extension of today's concept of ubiquitous computing, i.e. the integration of microprocessors into everyday objects. However, AmI will also be much more than this, as the AmI system should adapt to the user's needs and behavior.

IVT technologies and wireless technologies will play a fundamental role in developing the AmI vision to cope with the need for natural user interfaces and for ubiquitous communication. The former will enable the citizen to interact with the AmI and to control it in a natural and personalized way through voice and gestures. The latter will provide the underlying network and will also enable electronic devices to communicate with each other and with the user.

ISTAG identified a series of necessary characteristics that will permit the eventual societal acceptance of AmI. AmI should [36]:

- facilitate human contact;
- be orientated towards community and cultural enhancement;
- help to build knowledge and skills for work, better quality of work, citizenship and consumer choice;
- inspire trust and confidence;

- be consistent with long term sustainability both at personal, societal and environmental levels;
- be controllable by ordinary people - i.e. the 'off-switch' should be within reach (these technologies could very easily acquire an aspect of 'them controlling us'!).

Since the AmI vision is a long-term vision, as previously discussed the elaboration of scenarios is considered the most appropriate approach to identify the technology individual steps and challenges, and the bifurcations in technology development.

According to [36], "Scenarios are not straightforward extrapolations from the present, but offer provocative glimpses of futures that can (but need not) be realized. Each scenario has a script that is used to work out the key developments in technologies, society, economy, and markets necessary to arrive at the scenario. With the time-scale of significant changes in the ICT industry now shorter than one year, scenario planning provides one of the few structured ways to get an impression of the future."

Among different AmI scenarios which have been built up by several working groups in the US and in Europe, one very well known set of scenarios is the one by ISTAG. In [36] four scenarios have been described. We summarize them in Table 2.

4.2   Ambient Intelligence Technology Requirements

Using IVT technologies for communications means that interfaces have to be integrated with the real environment in such a way that user access to location-related content, including content concerning "virtual" locations, is as natural and intuitive as possible. Achieving this goal requires solutions to a number of technical and scientific problems. The IMSC of University of Southern California identified the following key issues [37]:

- current human-machine interfaces are in most cases inefficient, ineffective, one-way tools of communication, and there are currently no effective methodologies for managing, accessing, understanding and sharing integrated, distributed databases;
- real-time distribution and storage of multimedia content is complex and expensive;
- integration between multimedia technology and specific application contexts is poor, and this prevents the end user from making effective and productive use of these technologies;
- in areas such as medicine, commerce, communication and manufacturing industry, the lack of standard technologies and procedures implies that design, implementation and evaluation of modular multimedia solutions is costly, both in terms of time and of human resources.

In spite of this poor staring point, the AmI objective is to interface the environment with an individual who is communicating in natural language and through

Table 2. ISTAG scenarios for Ambient Intelligence.

Name	Environment	Characteristics	Key technologies
MARIA ("Road warrior")	An indoor environment of "augmented objects", controlled via a wrist-worn personal communications device	Support for highly mobile professionals	- Voice-commanded radio frequency transmitter/receiver supporting different wireless protocols (GSM, UMTS, Bluetooth, 802.11b) - Personalized 'Key of Keys' (e.g. biometric characteristics of the user) - Flexible "ad hoc" network management
DIMITRIOS ("Digital Me")	An outdoor environment where the user is supported by a Personal Digital Assistant (D-ME). The device engages in decision-making, virtual communications and exchanges of information with other D-ME devices	Support for the development and maintenance of social relations. Oriented to mass-market use	- A micro-sized, network-independent device with integrated data acquisition capabilities. The device should have the ability to recognize the emotions of a speaker and provide ad hoc wireless communications - Wireless LAN (e.g. Bluetooth) integrated with Wide Area Networks (WAN) (e.g. the cell-phone network)
CARMEN ("Traffic, sustainability, e-commerce")	An outdoor environment including an intelligent system for vehicle traffic management and goods delivery. The individual user is supported by a Personal Area Network	Support for personal and professional activities.	- Dynamically reconfigurable network devices - Multifunctional, reconfigurable sensor networks - Integration between Personal Area Networks (PAN), Vehicle Area Networks (VAN) e Wide Area Networks (WAN)
ANNETTE & SOLOMON ("Environment for social learning")	An indoor environment realized through interaction between far apart spaces	Support for study groups	- Identification of individual users, groups of users and objects - Networking between objects and users - Realization and management of virtual environments - Smart interfaces

gestures. Appendix accounts for some of the main research projects, already ongoing and planned, in the area of ambient intelligence.

A key problem is how to break through the barrier between the individual and the network, so realizing efficient person-machine interfaces. Achieving ever high degrees of integration between users and the AmI implies the existence of connections between electronic and telecommunications devices the user wears and the surrounding environment. An advanced network to allow the interconnection of a range of different wearable devices would constitute a Body Area Network (BAN) [38,39]. A BAN is a heterogeneous network which meets at least the following requirements:

- is capable of connecting both complete devices and individual device components in an easy and reconfigurable way;
- supports a range of different connection technologies - using both wired and wireless techniques including radio, infrared, and optical;
- is easy to use and to configure;
- supports a range of different classes of data (real time audio and video, Internet, etc.);
- allows the user to connect to the outside world;
- ensures security with respect to connections with the outside world.

As the BAN has to support interoperability between devices from different manufacturers it is necessary to define a common communications protocol. In this setting a BAN device could be either a complete piece of equipment (for example a cell-phone) or a single component (such as an earphone or a keyboard). BANs do not usually have to communicate over distances greater than 1-2 metres.

Personal Area Networks (PANs) are very similar to BANs. A PAN is defined as a short range (about 10 m) wireless network allowing the creation of ad-hoc, configurable connections to/from the outside world. While BANs provide local management of devices worn by the user PANs create a network around the user. Typical applications for a PAN include a range of personal computational services provided via laptops, PDAs and mobile phones. A PAN, might, for instance, provide access to a printer located in the vicinity of the user or Internet access via a local port.

BANs and PANs will be able to follow the user wherever he/she goes. Strong interactions between BAN/PAN and the environment will allow AmI to get accurate perception of the user. Strong and advanced forms of interaction between the PAN and the BAN will allow the user to receive richer feedback from the surrounding AmI.

The underpinning technologies required to construct the AmI landscape cover a broad range of ICT technologies. In the following we highlight some of the required technologies.

4.3   Enabling Technologies

The realization of the AmI vision requires the development of a number of enabling technologies: some of them are today under study, and a few are sketched in the following. We only concentrate below on interface technologies and mobile communications technologies. The former are focal to the development of IVT technologies. The latter aim to cope with the requirements of flexibility, scalability and efficient spectrum usage.

Vision technologies. Human vision is based on perception,[1] a complex hypothesis testing mechanism grounded on the two optical images provided by the eyes. Today's 3-D vision technologies are two: stereoscopy and holography. The former are more developed and are based on the parallax effect, exploited by the human vision system to perceive depth. Brain attempts to overlap the two different retinal images as much as possible by rotating the eyeballs towards a single focal point. The IMAX-3D [41] movie vision system is based on stereoscopy. The scene is filmed with two orthogonally polarized cameras located 6.5 cm apart (approximately the eyeball distance). The two images are then projected on one single movie theater screen. The spectator wears the IMAX-3D visor with polarized lenses matched to the two incoming waves.

Holography is a means to create an image without lenses. Differently from photography, in addition to the intensity of the light wave, holography memorizes its phase. Today it is still difficult to provide holographic moving images, so this technology is considered suitable for applications in a perspective beyond a ten year vision.

A number of other vision technologies are under study: this is one technology area in which some main breakthroughs will certainly happen: 3-D computer displays [42], miniaturized cameras for direct injection of images into the eye, etc. Another related technology is eyeball tracking.

Smart dust. "Smart dust" [43] is a cloud of tiny speckles, each one of millimeter dimension, of active silicon: the prototype under development at Berkeley University is shown in Figure 10. Mote senses, communicates and is self-powered. Each mote converts sunlight into electricity, locally elaborates information, localizes itself, both in absolute and relative to other particles, communicates with other ones within a few meters; furthermore, they jointly act as a coordinated array of sensors, a distributed computer, and as a communications network.

A similar technology is "smart painting", a random network of wall-painted computers studied at the MIT [44].

---

[1] "Perception is not determined simply by the stimulus patterns; rather it is a dynamic searching for the best interpretation of the available data. ... The senses do not give us a picture of the world directly; rather they provide evidence for checking hypotheses about what lies before us. Indeed, we may say that a perceived object *is* a hypothesis, suggested and tested by sensory data. The Necker cube is a pattern which contains no clues as to which of two alternative hypotheses is correct. The perceptual system entertains first one then the other hypothesis, and never comes to a conclusion, for there is no best answer. Sometimes the eye and brain comes to wrong conclusions, and then we suffer hallucinations or illusions." [40]

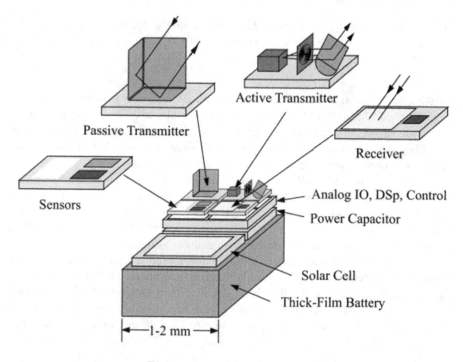

Fig. 10. Smart dust (from [43]).

Some features of smart dust are: free-space communication at optical frequencies, bit rates in the order of kilobits per second, power consumption of a few milliwatts, adoption of power management strategies, directional transmission within a narrow beam.

Radio reconfigurability. Radio reconfigurability (also known as "software radio") [45] is a set of techniques and rules aiming at allowing the radio system reconfiguration through software downloading, so that the system can readapt to different standards. The ideal software radio system is a single wireless platform configured in software to support any wireless standards for any use, in any place, at any time.

Expected impacts of software radio on system design are widespread: from base stations to terminals, and to network infrastructures, from standards definitions to spectrum regulation issues, etc. This will be a fundamental technology to reach the objective of flexibility at equipment level, the objective of scalability at system level, and that of optimum use of frequency spectrum.

Smart antennas. Smart antennas aim at dynamically controlling interference produced into the system and external interference captured from it [46]. In transmission a smart antenna ideally radiates power only towards the desired user. In reception it ideally captures power only from the desired user. These antennas are now already considered in cellular system base station designs for inclusion as an option within 3G systems. In the future, their use will be extremely widespread, as spectrum efficiency will be one major target in system design.

Stratospheric platforms. Stratospheric platforms, also known as High Altitude Local Orbiters (HALOs), are among possible new system developments to provide direct and immediate WAN access to telecommunications resources [47]. They are quasi-geostationary platforms located at an altitude around 25 km above ground operating at millimeter wave frequencies.

HALOs can provide cellular coverage in line-of-sight. Their very large elevation angles (more than 60 deg) promise visibility much better than that achievable with geostationary satellite systems; the usage of millimeter wave frequencies can allow use of very large bandwidths, which can result in system capacity much larger than achievable with cellular systems.

Main expected applications are last mile interactive multimedia services distribution to low cost rooftop terminals. Each stratospheric platform acts as the highest tower in town, providing high density, high capacity, high speed services, with low power requirements, and no latency, mainly to metropolitan and suburban areas. On-board switching is adopted for direct connection between subscribers within the same platform coverage area. Traffic for subscribers outside the stratospheric platform coverage area will be routed through ground stations and/or by means of inter-platform links.

Ultra wide band communications. Ultra wide band (UWB) communication (also known as "Impulse radio") [48] is a technique to spread the signal bandwidth over an exceptionally large bandwidth (well beyond those of 3G CDMA systems), starting from frequencies close to baseband and with cancellation of the continuous wave. Through the use of extremely narrow pulses (duration of a few nanoseconds) the r.f. bandwidth is in excess of 1 GHz. UWB allows large capacity multiple access, very large values of individual bit rates (e.g. 100 Mbit/s), and very low power spectral densities. Furthermore, the intrinsic diversity provided by the ultra wide bandwidth, combined with the use of RAKE receiver structures, allows multipath fading immunity and better and stable transmission quality. Finally, the usage of low frequencies provides improved penetration of radio waves through walls.

UWB will allow spectrum coexistence of different signals for different services overcoming the present problem of inefficient use of large portions of spectrum, permanently and exclusively allocated and poorly used. Main expected applications are indoor interactive multimedia communications, localization and distance measurements. UWB is therefore one possible technology candidate to wirelessly distribute multimedia services provided through last mile technologies, such as e.g. HALOs. Presently, the Federal Communications Commission in the U.S. started the standardization procedure to allow commercial use of UWB [49].

5   Conclusions

This tutorial papers aimed at highlighting some main issues related to the development of virtual technologies, wireless communications technologies and ambient intelligent technologies. These three sectors will be central to the development of a new technology perspective which is expected to materialize in the 2010s.

References

1. Riva, G., Galimberti, C.: The psychology of cyberspace: a socio-cognitive framework to computer mediated communication. New Ideas in Psychology **15** (1997) 141–158
2. Rockart, J.F., DeLong, D.W.: Executive support systems: the emergence of top management computer use. Dow-Jones-Irwin, Homewood, IL, USA (1988)
3. Sheridan, T.B.: Telerobotics. Automatica **25** (1989) 487–507
4. Durlach, N.: The potential of teleoperation for entertainment and education. Presence, Teleoperators, and Virtual Environments **6** (1997) 350–351
5. Biocca, F., R.Levy, M.: Applicazioni di realtà virtuale nell'ambito della comunicazione. In: La comunicazione virtuale, C. Galimberti and G. Riva (eds.). Guerini e Associati, Milano (1997) 79–110
6. Riva, G.: From technology to communication: psycho-social issues in developing virtual environments. Journal Visual Languages and Computing **10** (1999) 87–97
7. Riva, G.: Virtual reality as a communication tool: a socio cognitive analysis. Presence, Teleoperators, and Virtual Environments **8** (1999) 460–466
8. Steuer, J.: Definire la realtà virtuale: le dimensioni che determinano la telepresenza. In: La comunicazione virtuale, C. Galimberti and G. Riva (eds.). Guerini e Associati, Milano (1997) 55–78
9. Bricken, W.: Virtual reality: directions of growth. Technical Report R-90-1, HITL, University of Washington, Seattle, WA (1990)
10. Schloerb, D.: A quantitative measure of telepresence. Presence, Teleoperators, and Virtual Environments **4** (1995) 64–80
11. Slater, M., Wilbur, S.: A framework for immersive virtual environments (FIVE): speculations on the role of presence in virtual environments. Presence, Teleoperators, and Virtual Environments **6** (1997) 603–616
12. Hoffman, H., Hollander, A., Schroder, K., Rousseau, S., Furness, T.: Physically touching and tasting virtual objects enhances the realism of virtual experiences. Virtual Reality **3** (1998) 226–234
13. Sastry, L., Boyd, D.R.S.: Virtual environments for engineering applications. Virtual Reality **3** (1998) 235–224
14. Zahoric, P., Jenison, R.L.: Presence as being in the world. Presence, Teleoperators, and Virtual Environments **7** (1998) 78–89
15. Mantovani, G., Riva, G.: 'Real' presence: how different anthologies generate different criteria for presence, telepresence, and virtual presence. Presence, Teleoperators, and Virtual Environments **8** (1999) 538–548
16. Cole, M.: Cultural psychology: a once and future discipline. Harvard University Press, Cambridge, MA, USA (1996)
17. Stone, A.R.: The war of desire and technology at the close of the Mechanical Age. MIT Press, Cambridge, MA, USA (1996)

148   F. Davide et al.

18. Churchill, E.F., Snowdon, D.: Collaborative virtual environments: an introductory review of issues and systems. Virtual Reality **3** (1998) 3–15
19. Grundin, J., Poltrock, S.E.: Groupware and workflow: experiences, state-of-the-art and future trends. In: Conference on Human Factors in Computing Systems, Vancouver, British Columbia, Canada (1996)
20. Benford, S., Greenhalgh, C., Reynard, G., Brown, C., Koleva, B.: Understanding and constructing shared spaces with mixed reality boundaries. ACM Transactions on Computer-Human Interaction **5** (1998) 185–223
21. Galimberti, C., Riva, G.: L'interazione virtuale: nuove tecnologie e processi comunicativi. In: La comunicazione virtuale, C. Galimberti and G. Riva (eds.). Guerini e Associati, Milano (1997) 15–53
22. Bersot, O., El-Guedj, P.O., Godéreaux, C., Nugues, P.: A conversational agent to help navigation and collaboration in virtual worlds. Virtual Reality **3** (1998) 78–82
23. Riva, G., Bacchetta, M., et al.: The use of PC based VR in clinical medicine: the VREPAR projects. Technology and Health Care **7** (1999) 261–269
24. Beach, L.R.: Image Theory: Decision Making in Personal and Organizational Contexts. John Wiley & Sons (1990)
25. Schmidt, K.: Riding a tiger, or computer supported cooperative work. In: Second European Conference on Computer Supported Cooperative, WorkL. Bannon, M. Robinson and K. Schmidt (eds.). Guerini e Associati, Dordrecht, Kluwer (1991) 55–78
26. Weick, K.E.: Sensemaking in Organizations. Sage (1995)
27. Mantovani, G.: Comunicazione e identità: dalle situazioni quotidiane agli ambienti virtuali. Il Mulino, Bologna (1995)
28. Suchman, L.: Constructing shared workspaces. In: Cognition and communication at work, Y. Engeström and D. Middleton (eds.). Cambridge University Press, New York, NY (1996)
29. Kaaranen, H., et al.: UMTS Networks. John Wiley & Sons (2001)
30. WWRF: The book of visions 2001. Technical report, Wireless World Research Forum (WWRF) (2001)
31. EICTA: Position on priorities for information and communications technology r&d collaborative programmes in the european research areas. Technical report, EICTA, Brussels, B (2001)
32. Christopoulos, C.: Mobile augmented reality (MAR) and virtual reality. Technical report, Wireless World Research Forum (WWRF) (2001)
33. Baez, M.G.O., Danielsson, P.: Employing electrical field sensing for detecting static thumb position using the finger-joint gesture keypad input paradigm. In 173-174, ed.: The Fourth International Symposium on Wearable Computers, IEEE. (2000)
34. Anoto Group: Anoto website. http://www.anoto.com/ (2002)
35. Saracco, R.: The future of communications: managing what, managing how? In: Internationel Symposium on Integrated Network Management, Seattle, Washington, USA (2001)
36. Ducatel, K., Bogdanowicz, M., Scapolo, F., Leijten, J., Burgelma, J.C.: Scenarios for ambient intelligence in 2010. ISTAG 2001, Final Report, IPTS-Seville, (http://www.cordis.lu/ist/istag.htm) (2001)
37. Nikias, M., et al.: IMSC year four annual report. Technical Report 1-2, University of Southern California, Integrated Media Systems Center, Los Angeles, CA (2000)
38. Ditlea, S.: The PC goes ready to wear. IEEE Spectrum **37** (2000) 34–39
39. Van Dam, K., et al.: From PAN to BAN: Why body area networks. In: Proceedings of the Wireless World Research Forum (WWRF) meeting, Helsinki, Finland (2001) 10–11

40. Drascic, D., Milgram, P.: Perceptual Issues in Augmented Reality. In: Stereoscopic Displays and Virtual Reality Systems III. Volume 2653. SPIE Press (1996) 123–134
41. IMAX Corporation: Imax website. http://www.imax.com/ (2002)
42. Dipert, B.: Display enhancements accept no compromises. EDN Magazine (2000) 47–56
43. Kahn, J.M., Katz, R.H., Pister, K.: Next century challenges: Mobile networking for 'Smart Dust'. In: ACM International Conference on Mobile Computing and Networking, Seattle, WA, USA (1999)
44. Massachusetts Institute of Technology: Media Laboratory website. http://www.media.mit.edu/ (2002)
45. Mitola, J.: Software Radio Architecture: Object-Oriented Approaches to Wireless Systems Engineering. John Wiley & Sons (2000)
46. Giuliano, R., Mazzenga, F., Vatalaro, F.: Smart cell sectorization for third generation CDMA systems. Wireless Communications & Mobile Computing, to be published (2002)
47. Djuknic, G., Freidenfelds, J., Okunev, Y.: Establishing wireless communications services via high-altitude aeronautical platforms: a concept whose time has come? IEEE Communications Magazine 35 (1997) 128–135
48. Win, M.Z., Scholtz, R.A.: Ultra-wide bandwidth time-hopping spread-spectrum impulse radio for wireless multiple-access communications. IEEE Transactions on Communications 48 (2000) 679–689
49. FCC: New public safety applications and broadband internet access among uses envisioned by FCC authorization of ultra-wideband technology. Technical report, Federal Communications Commission (2002)
50. Hagsand, O.: Interactive multi-user VEs in the DIVE system. IEEE Multimedia 3 (1996) 30–39
51. Normand, V., et al.: The coven project: exploring applicative, technical and usage dimensions of collaborative virtual environments. Presence, Teleoperators and Virtual Environment 8 (1999) 218–336
52. Greenhaigh, C., Benford, S.D.: Supporting rich and dynamic communication in large-scale collaborative virtual environments. Presence, Teleoperators, and Virtual Environments 8 (1999) 14–35
53. University of Washington: Human Interface Technology Laboratory website. www.hitl.washington.edu/ (2002)

Appendix: Research on Virtual Technologies

European VICE Projects

DIstributed Virtual Environment (DIVE). DIVE is a virtual immersive cooperative environment (VICE) created by the Swedish Institute of Computer Science. The goal is to build a modular environment which can be easily adapted to specific situations [50]. The environment consists of a structured family of objects that allows a programmer to separate different phases in data processing and to manage complex interactions without need for specific control routines. This makes it relatively easy to create applications for specific environments. As a result programmers created a number of environments sharing the same structure and interface. One of these is VR-VIBE, a VICE designed to facilitate literature search and the display of search results [18]. In this environment information is represented as a set of colored polygons whose color depends on the kind of search performed by users. When users have their search completed, they can navigate through the environment. When they wish to consult complete literature information they interact with individual polygons. Users who have worked with VR-VIBE generally find it satisfactory. However, test results show that the shared use of the environment by multiple users can lead to difficulties. One of the most critical of these is the definition of shared search criteria. The attempt to solve this problem has led to the creation of the COllaborative Virtual ENvironment (COVEN), with the goal to design and implement new DIVE tools able to facilitate multi-user interaction. COVEN is devoting special attention to the features provided by avatars, to ways in which users can modify their viewpoint to adapt to other actors, and to the implementation of voice-based communications [51].

Model Architecture and System for Spatial Interaction in Virtual Environments (MASSIVE). MASSIVE has been implemented by researchers from the University of Nottingham, UK [20]. Like DIVE, MASSIVE allows adopting a scalable approach to environment implementation. In particular, the system makes it possible to build separate "worlds" connected by "foyers". What differentiates MASSIVE from DIVE is the decision to use a spatial interaction model based on the concepts of "aura", "focus" and "nimbus". Aura denotes an area of space surrounding each of the actors and objects within the virtual environment. Interactions take place within auras. Two individuals can interact only when their auras collide. The concepts of focus and nimbus make it possible to define actors' awareness during the interaction. To be more specific, focus allows to communicate one actor's level of awareness to another actor or to an object. Nimbus allows an actor to modify the way in which he/she is perceived when enters another actor's aura. The spatial model has been recently extended [52] by introducing "third-party objects" with the ability to alter actors' levels of awareness. In this way, for example, a designer can create isolated rooms which are impenetrable to external auras. The use of a spatial interaction model should be considered as a step forward towards the implementation of a genuine collaborative environment. By modifying the aura, the focus and the nimbus, actors can define "zones of interest". This makes it possible to modularize group activities and to use MASSIVE in a range of situations involving collaborative work.

MAintenancE System based on Telepresence for Remote Operators (MAESTRO). This project, coordinated by Thomson CSF/LCR, has developed a distributed system based on augmented reality for maintenance and repair and for training of maintenance workers. The user interacts with the malfunctioning equipment via a video-screen where real images are merged with multimedia information and scenes describing the equipment under test. The MAESTRO platform integrates a number of different technologies: a 3-D tracker (to localize bodies in space), special navigation and manipulation peripherals, 3-D modelling, voice recognition, synchronized rendering of hybrid environments (video sources, audio, 3-D models, images, documents), multimedia streaming over networks with guaranteed QoS.

Avatar based conferencing in virtual worlds for business purposes (AVATAR CONFERENCE). AVATAR CONFERENCE is a toolkit for the set-up and administration of virtual online conferences in which users are represented as avatars. The system provides means for spontaneous and intuitive communications between partners, through a multimedia multi-user real-time communication, speech and voice recognition facilities, online translation services, user representation within virtual worlds, multi-user 3D-manipulation and whiteboard, application sharing facilities on PC.

European Ambient Interaction Projects

Theatre of Work Enabling Relationships (TOWER). TOWER is a e-Work project aiming at developing a new cooperative environment and workspace to enable new ways of working over a distance for synchronous, as well as asynchronous working modes. It intends to augment existing groupware systems through sensors and agents, which recognize awareness information.

A new paradigm for high-quality mixing of real and virtual (ORIGA-MI). This European research project is about the development of new methodologies for the seamless mixing of real and virtual elements. This is applied to cinema and TV production, including both interactive and diffusive services. A digital analysis of reality with advanced real-time feedback to control movement and position. The project involves three European universities, three big production companies for cinema, TV and advertisement (BBC, Framestore, China Town) and an Italian company that develops software for special effects in cinema applications.

Advanced Distributed for sensor network Architecture (ADA). The project deals with the sensing devices and data collection parts of the environmental monitoring systems. The project aims in to define an architecture for supporting data transmission combined with the promising innovations in the areas of intelligent sensors and intelligent terminals.

Virtual Immersive COMmunications (VICOM). (funding approval from the Italian research ministry in progress) VICOM aims to study and develop techniques, protocols and applications leading to the implementation and evaluation of two major demonstrators in the field of VIT. The demonstrators have been chosen to represent complementary needs in terms both of applications scenarios and technologies (infrastructure platforms, development systems, support terminals). With this goal in mind the demonstrators will deal respectively with mobility in immersive environments and with immersive tele-training. Both these environments will require the study of enabling technologies - in particular mobile and fixed telecommunications networks and distributed information processing.

Research at MIT MediaLab

The Massachusetts Institute of Technology (MIT) Medialab [44] hosts several research groups that are investigating on specific aspects of AmI. Some main activities of these groups are referred below:

Affective Computing. The Affective Computing research group aims to bridge the gap between computational systems and human emotions. Their research addresses machine recognition and modeling of human emotional expression, machine learning of human preferences as communicated by user affect, intelligent computer handling of human emotions, computer communication of affective information between people, affective expression in machines and computational toys, emotion modeling for intelligent machine behavior; tools to help develop human social-emotional skills, and new sensors and devices to help gather, communicate, and express emotional information. As an example, one project ("Recognizing Affect in Speech") aims at building computational models for the automatic recognition of affective expressions in speech.

Context-Aware Computing. The Context-Aware Computing group is interested in using "context knowledge" such as what we do, what we have done, where we are and how we feel about it, in order to interact with the environment around us. By understanding and using the contextual reality of a situation, they aim at redefining person-computer interaction. One example is a project ("Disruption") attempting to make a computer intelligent enough to select the appropriate sensorial modality based on the user's preferred perceptual channel. This project investigates which modality is the most efficient and at the same time, the least disruptive. Five interruption modalities are studied: heat, smell, sound, vibration, and light. An important observation is that subjects are affected differently by each of the modalities. Subjects' reactions depended on their previous exposure to a specific modality. They will build a notification system that finds a user's preference of interruptions, adapts to those preferences, and uses them to interrupt the user in an incremental way. The system will then be a self-adaptive interface. Another example is a system ("Eye-aRe Eye-aRe") designed to detect and communicate the intentional information conveyed in eye movement. A glasses-mounted wireless device stores and transfers information based on users' eye motion and external IR devices, thus promoting an enriched experience with their environment. The system measures eye motion and utilizes this as an implicit input channel to a sensor system and computer. Eye motion detection can be used to recognize a users' gaze. When the person's eyes are fixed the system infers that he is paying attention to something in the environment and then it attempts to facilitate an exchange of information in either direction on the user's behalf.

Tangible Media. We live between two worlds: our physical environment and the cyberspace. Tangible Media group's focus is on the design of seamless interfaces between humans, digital information, and the physical environment. People have developed sophisticated skills for sensing and manipulating our physical environments. However, most of these skills are not employed by traditional graphical user interfaces. The Tangible Media group is designing a variety of "tangible interfaces" based on these skills by giving physical form to digital information, seamlessly coupling the two worlds. In this framework the "ComTouch" project explores interpersonal communication by means of haptic technology. Touch as an augmentation to existing communication may provide and enhance existing audio and video media. ComTouch is a hand-held device that allows the squeeze under each finger to be represented as vibration. This haptic device aims at enabling users to transmit thoughts, feelings, and concepts to each other remotely. The form factor is designed to fit on the back of a cellular phone; as users talk on the cell phone, they can squeeze and transmit a representation of their touch. Another project ("inTouch") explores new forms of interpersonal communication through touch. Force-feedback technology is employed to create the illusion that people, separated by distance, are interacting with a shared physical object. The "shared" object provides a haptic link between geographically distributed users, opening up a channel for physical expression over distance.

Research at Washington University HIT Lab

The Human Interface Technology (HIT) Lab at Washington University is engaged in a number of projects on Mixed Reality [53]. In the follow we present two of them.

Shared Space. The Shared Space interface intends demonstrating how augmented reality can enhance both face-to-face and remote collaboration. For face-to-face collaboration, it allows users to see each other and the real world at the same time as 3-D virtual images between them, supporting natural communication between users and intuitive manipulation of the virtual objects. For remote collaboration, the system allows life-sized live virtual video images of remote user to be overlaid on the local real environment, supporting spatial cues and removing the need to be physically present at a desktop machine to conference.

Multimodal Interfaces. This project involves the development of software libraries for incorporating multimodal input into human computer interfaces. These libraries combine natural language and artificial intelligence techniques to allow human computer interaction with an intuitive mix of voice, gesture, speech, gaze and body motion. Interface designers will be able to use this software for both high and low level understanding of multimodal input and generation of the appropriate response.

A Tutorial on Optical Networks

George N. Rouskas and Harry G. Perros

Department of Computer Science, North Carolina State University, Raleigh, NC, USA
rouskas,hp@csc.ncsu.edu

Abstract. In this half-day tutorial, we present the current state-of-the-art in optical networks. We begin by discussing the various optical devices used in optical networks. Then, we present wavelength-routed networks, which is currently the dominant architecture for optical networks. We discuss wavelength allocation policies, calculation of call blocking probabilities, and network optimization techniques. Subsequently, we focus on the various protocols that have been proposed for wavelength-routed networks. Specifically, we present a framework for IP over optical networks, MPLS, LDP, CR-LDP, and GMPLS. Next, we discuss optical packet switching and optical burst switching, two new emerging and highly promising technologies.

1   Introduction

Over the last few years we have witnessed a wide deployment of point-to-point wavelength division multiplexing (WDM) transmission technology in the Internet infrastructure. The corresponding massive increase in network bandwidth due to WDM has heightened the need for faster switching at the core of the network. At the same time, there has been a growing effort to enhance the Internet Protocol (IP) to support traffic engineering [1,2] as well as different levels of Quality of Service (QoS) [3]. Label Switching Routers (LSRs) running Multi-Protocol Label Switching (MPLS) [4,5] are being deployed to address the issues of faster switching, QoS support, and traffic engineering. On one hand, label switching simplifies the forwarding function, thereby making it possible to operate at higher data rates. On the other hand, MPLS enables the Internet architecture, built upon the connectionless Internet Protocol, to behave in a connection-oriented fashion that is more conducive to supporting QoS and traffic engineering.

The rapid advancement and evolution of optical technologies makes it possible to move beyond point-to-point WDM transmission systems to an all-optical backbone network that can take full advantage of the available bandwidth. Such a network consists of a number of optical cross-connects (OXCs) arranged in some arbitrary topology, and its main function is to provide interconnection to a number of IP/MPLS subnetworks. Each OXC can switch the optical signal coming in on a wavelength of an input fiber link to the same wavelength in an output fiber link. The OXC may also be equipped with converters that permit it to switch the optical signal on an incoming wavelength of an input fiber to

E. Gregori et al. (Eds.): Networking 2002 Tutorials, LNCS 2497, pp. 155–193, 2002.

some other wavelength on an output fiber link. The main mechanism of transport in such a network is the lightpath (also referred to as λ-channel), an optical communication channel established over the network of OXCs which may span a number of fiber links (physical hops). If no wavelength converters are used, a lightpath is associated with the same wavelength on each hop. This is the well-known wavelength continuity constraint. Using converters, a different wavelength on each hop may be used to create a lightpath. Thus, a lightpath is an end-to-end optical connection established between two subnetworks attached to the optical backbone.

Currently, there is tremendous interest within both the industry and the research community in optical networks in which OXCs provide the switching functionality. The Internet Engineering Task Force (IETF) is investigating the use of Generalized MPLS (GMPLS) [6] and related signaling protocols to set up and tear down lightpaths. GMPLS is an extension of MPLS that supports multiple types of switching, including switching based on wavelengths usually referred to as Multi-Protocol Lambda Switching (MPλS). With GMPLS, the OXC backbone and the IP/MPLS subnetworks will share common functionality in the control plane, making it possible to seamlessly integrate all-optical networks within the overall Internet infrastructure. Also, the Optical Domain Service Interconnection (ODSI) initiative (which has completed its work) and the Optical Internetworking Forum (OIF) are concerned with the interface between an IP/MPLS subnetwork and the OXC to which it is attached as well as the interface between OXCs, and have several activities to address MPLS over WDM issues [7]. Optical networks have also been the subject of extensive research [8] investigating issues such as virtual topology design [9,10], call blocking performance [11,12], protection and restoration [13,14], routing algorithms and wavelength allocation policies [15,16,17], and the effect of wavelength conversion [18,19,20], among others.

The tutorial is organized as follows. Section 2 introduces the basic elements of the optical network architecture, and Section 3 presents the routing and wavelength assignment problem, the fundamental control problem in optical networks. Section 4 discusses standardization activities under way for optical networks, with an emphasis on control plane issues. Section 5 discusses a framework for IP over optical networks, MPLS, the signaling protocols LDP and CR-LDP, and GMPLS. Section 6 describes optical packet switching, and finally, Section 7 describes an emerging technology, optical burst switching.

2   Wavelength Routing Network Architecture

The architecture for wide-area WDM networks that is widely expected to form the basis for a future all-optical infrastructure is built on the concept of *wavelength routing*. A wavelength routing network, shown in Figure 1, consists of *optical cross-connects (OXCs)* connected by a set of fiber links to form an arbitrary mesh topology. The services that a wavelength routed network offers to attached client subnetworks are in the form of *logical* connections implemented using *lightpaths*. Lightpaths are clear optical paths which may traverse a number of fiber links in the optical network. Information transmitted on a lightpath does not undergo any conversion to and from electrical form within the optical network, and thus, the architecture of the OXCs can be very simple because they do not need to do any signal processing. Furthermore, since a lightpath behaves as a literally transparent "clear channel" between the source and destination subnetwork, there is nothing in the signal path to limit the throughput of the fibers.

The OXCs provide the switching and routing functions for supporting the logical data connections between client subnetworks. An OXC takes in an optical signal at each of the wavelengths at an input port, and can switch it to a particular output port, independent of the other wavelengths. An OXC with N input and N output ports capable of handling W wavelengths per port can be thought of as W independent $N \times N$ optical switches. These switches have to be preceded by a wavelength demultiplexer and followed by a wavelength multiplexer to implement an OXC, as shown in Figure 2. Thus, an OXC can cross-connect the different wavelengths from the input to the output, where the connection pattern of each wavelength is independent of the others. By appropriately configuring the OXCs along the physical path, logical connections (lightpaths) may be established between any pair of subnetworks.

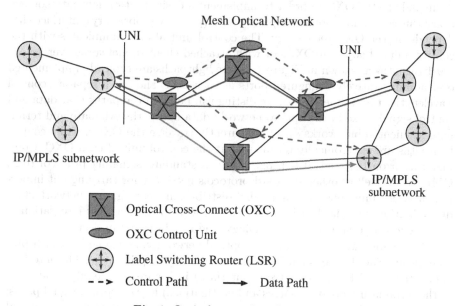

Fig. 1. Optical network architecture

As Figure 1 illustrates, each OXC has an associated *electronic* control unit attached to one of its input/output ports. The control unit is responsible for control and management functions related to setting up and tearing down lightpaths;

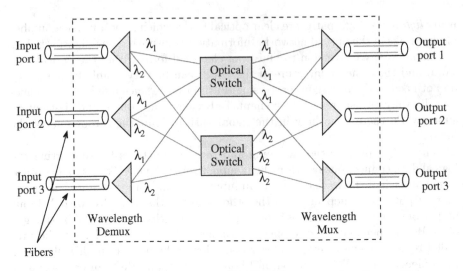

Fig. 2. 3 × 3 optical cross-connect (OXC) with two wavelengths per fiber

these functions are discussed in detail in Section 4. In particular, the control unit communicates directly with its OXC, and is responsible for issuing configuration commands to the OXC in order to implement a desired set of lightpath connections; this communication takes place over a (possibly proprietary) interface that depends on the OXC technology. The control unit also communicates with the control units of adjacent OXCs or with attached client subnetworks over *single-hop* lightpaths as shown in Figure 1. These lightpaths are typically implemented over administratively configured ports at each OXC and use a separate control wavelength at each fiber. Thus, we distinguish between the paths that data and control signals take in the optical network: data lightpaths originate and terminate at client subnetworks and transparently traverse the OXCs, while control lightpaths are electronically terminated at the control unit of each OXC. Communication on the control lightpaths uses a standard signaling protocol (e.g., GMPLS), as well as other standard protocols necessary for carrying out important network functions including label distribution, routing, and network state dissemination. Standardization efforts are crucial to the seamless integration of multi-vendor optical network technology, and are discussed in Section 4.

Client subnetworks attach to the optical network via edge nodes which provide the interface between non-optical devices and the optical core. This interface is denoted as UNI (user-to-network interface) in Figure 1. The edge nodes act as the terminating points (sources and destinations) for the optical signal paths; the communication paths may continue outside the optical network in electrical form. In Figure 1, only the label switching routers (LSRs) of the two IP/MPLS subnetworks which are directly attached to an OXC implement the UNI and may originate or terminate lightpaths. For the remainder of this chapter we will make the assumption that client subnetworks run the IP/MPLS protocols. This assumption reflects the IP-centric nature of the emerging control architecture for optical networks [21]. However, edge nodes supporting any network technology (including ATM switches and SONET/SDH devices) may connect to the optical network as long as an appropriate UNI is defined and implemented.

In [22,23], the concept of a lightpath was generalized into that of a *light-tree*, which, like a lightpath, is a clear channel originating at a given source node and implemented with a single wavelength. But unlike a lightpath, a light-tree has multiple destination nodes, hence it is a point-to-multipoint channel. The physical links implementing a light-tree form a tree, rooted at the source node, rather than a path in the physical topology, hence the name. Light-trees may be implemented by employing optical devices known as *power splitters* [24] at the OXCs. A power splitter has the ability to split an incoming signal, arriving at some wavelength λ, into up to m outgoing signals, $m \geq 2$; m is referred to as the *fanout* of the power splitter. Each of these m signals is then independently switched to a different output port of the OXC. Note that due to the splitting operation and associated losses, the optical signals resulting from the splitting of the original incoming signal must be amplified before leaving the OXC. Also, to ensure the quality of each outgoing signal, the fanout m of the power splitter may have to be limited to a small number. If the OXC is also capable of wavelength conversion, each of the m outgoing signal may be shifted, independently of the others, to a wavelength different than the incoming wavelength λ. Otherwise, all m outgoing signals must be on the same wavelength λ.

An attractive feature of light-trees is the inherent capability for performing multicasting in the optical domain (as opposed to performing multicasting at a higher layer, e.g., the network layer, which requires electro-optic conversion). Such wavelength routed light-trees are useful for transporting high-bandwidth, real-time applications such as high-definition TV (HDTV). Therefore, OXCs equipped with power splitters will be referred to as *multicast-capable* OXCs (MC-OXCs). Note that, just like with converter devices, incorporating power splitters within an OXC is expected to increase the network cost because of the need for power amplification and the difficulty of fabrication.

3 Routing and Wavelength Assignment (RWA)

A unique feature of optical WDM networks is the tight coupling between routing and wavelength selection. As can be seen in Figure 1, a lightpath is implemented by selecting a path of physical links between the source and destination edge nodes, and reserving a particular wavelength on each of these links for the lightpath. Thus, in establishing an optical connection we must deal with both routing (selecting a suitable path) and wavelength assignment (allocating an available wavelength for the connection). The resulting problem is referred to as the *routing and wavelength assignment (RWA)* problem [17], and is significantly more difficult than the routing problem in electronic networks. The additional complexity arises from the fact that routing and wavelength assignment are subject to the following two constraints:

1. *Wavelength continuity constraint:* a lightpath must use the same wavelength on all the links along its path from source to destination edge node.
2. *Distinct wavelength constraint:* all lightpaths using the same link (fiber) must be allocated distinct wavelengths.

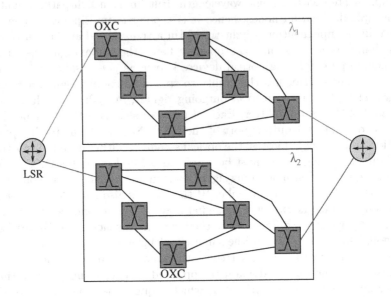

Fig. 3. The RWA problem with two wavelengths per fiber

The RWA problem in optical networks is illustrated in Figure 3, where it is assumed that each fiber supports two wavelengths. The effect of the wavelength continuity constraint is represented by replicating the network into as many copies as the number of wavelengths (in this case, two). If wavelength i is selected for a lightpath, the source and destination edge node communicate over the i-th copy of the network. Thus, finding a path for a connection may potentially involve solving W routing problems for a network with W wavelengths, one for each copy of the network.

The wavelength continuity constraint may be relaxed if the OXCs are equipped with *wavelength converters* [18]. A wavelength converter is a single input/output device that converts the wavelength of an optical signal arriving at its input port to a different wavelength as the signal departs from its output port, but otherwise leaves the optical signal unchanged. In OXCs without a wavelength conversion capability, an incoming signal at port p_i on wavelength λ can be optically switched to any port p_j, but must leave the OXC on the same wavelength λ. With wavelength converters, this signal could be optically switched to any port p_j on some other wavelength λ'. That is, wavelength conversion allows a lightpath to use different wavelengths along different physical links.

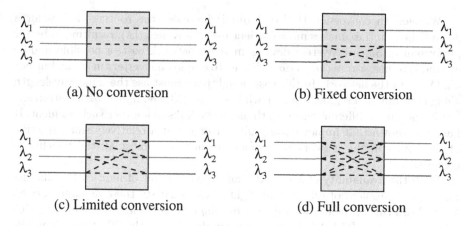

Fig. 4. Wavelength conversion

Different levels of wavelength conversion capability are possible. Figure 4 illustrates the differences for a single input and single output port situation; the case for multiple ports is more complicated but similar. *Full wavelength conversion* capability implies that any input wavelength may be converted to any other wavelength. *Limited wavelength conversion* [25] denotes that each input wavelength may be converted to any of a specific set of wavelengths, which is not the set of all wavelengths for at least one input wavelength. A special case of this is *fixed wavelength conversion*, where each input wavelength can be converted to exactly one other wavelength. If each wavelength is "converted" only to itself, then we have no conversion.

The advantage of full wavelength conversion is that it removes the wavelength continuity constraint, making it possible to establish a lightpath as long as each link along the path from source to destination has a free wavelength (which could be different for different links). As a result, the RWA problem reduces to the classical routing problem, that is, finding a suitable path for each connection in the network. Referring to Figure 3, full wavelength conversion collapses the W copies of the network into a single copy on which the routing problem is solved. On the other hand, with limited conversion, the RWA problem becomes more complex than with no conversion. To see why, note that employing limited conversion at the OXCs introduces links between *some* of the network copies of Figure 3. For example, if wavelength λ_1 can be converted to wavelength λ_2 but not to wavelength λ_3, then links must be introduced from each OXC in copy 1 of the network to the corresponding OXC in copy 2, but not to the corresponding OXC in copy 3. When selecting a path for the connection, at each OXC there is the option of remaining at the same network copy or moving to another one, depending on the conversion capability of the OXC. Since the number of alternatives increases exponentially with the number of OXCs that need to be traversed, the complexity of the RWA problem increases accordingly.

Wavelength conversion (full or limited) increases the routing choices for a given lightpath (i.e., makes more efficient use of wavelengths), resulting in better performance. Since converter devices increase network cost, a possible middle ground is to use *sparse conversion*, that is, to employ converters in some, but not all, OXCs in the network. In this case, a lightpath must use the same wavelength along each link in a segment of its path between OXCs equipped with converters, but it may use a different wavelength along the links of another such segment. It has been shown that implementing full conversion at a relatively small fraction of the OXCs in the network is sufficient to achieve almost all the benefits of conversion [11,19].

With the availability of MC-OXCs and the existence of multicast traffic demands, the problem of establishing light-trees to satisfy these demands arises. We will call this problem the *multicast routing and wavelength assignment (MC-RWA)* problem. MC-RWA bears many similarities to the RWA problem discussed above. Specifically, the tight coupling between routing and wavelength assignment remains, and even becomes stronger: in the absence of wavelength conversion the same wavelength must be used by the multicast connection not just along the links of a single path but along all the links of the light-tree. Since the construction of optimal trees for routing multicast connections is by itself a hard problem [26], the combined MC-RWA problem becomes even harder.

Routing and wavelength assignment is the fundamental control problem in optical WDM networks. Since the performance of a network depends not only on its physical resources (e.g., OXCs, converters, fibers links, number of wavelengths per fiber, etc.) but also on how it is controlled, the objective of an RWA algorithm is to achieve the best possible performance within the limits of physical constraints. The RWA (and MC-RWA) problem can be cast in numerous forms. The different variants of the problem, however, can be classified under one of two broad versions: a static RWA, whereby the traffic requirements are known in advance, and a dynamic RWA, in which a sequence of lightpath requests arrive in some random fashion. The static RWA problem arises naturally in the design and capacity planning phase of architecting an optical network, and is discussed in Section 3.1. The dynamic RWA problem is encountered during the real-time network operation phase and involves the dynamic provisioning of lightpaths; this issue is addressed in Section 3.2.

3.1   Static RWA

If the traffic patterns in the network are reasonably well-known in advance and any traffic variations take place over long time scales, the most effective technique for establishing optical connections (lightpaths) between client subnetworks is by formulating and solving a static RWA problem. Therefore, static RWA is appropriate for provisioning a set of semipermanent connections. Since these connections are assumed to remain in place for relatively long periods of time, it is worthwhile to attempt to optimize the way in which network resources (e.g., physical links and wavelengths) are assigned to each connection, even though optimization may require a considerable computational effort. Because off-line algorithms have knowledge of the entire set of demands (as opposed to on-line algorithms that have no knowledge of future demands), they make more efficient use of network resources and project a lower overall capacity requirement.

Physical Topology Design. In this phase the network operator has a demand forecast and must decide on a topology to connect client subnetworks through OXCs. This step includes the sizing of links (e.g., determining the number of wavelength channels and the capacity of each channel) and OXCs (e.g, determining the number of ports), as well as the placement of resources such as amplifiers, wavelength converters, and power splitters. Moreover, to deal with link or OXC failures, it is desirable to ensure that there are at least two (or three) paths between any pair of OXCs in the network, i.e., that the graph corresponding to the physical topology of the optical network is two- or three-connected. Often, geographical or administrative considerations may impose further constraints on the physical topology.

If a network does not already exist, the physical topology must be designed from scratch. Obviously, the outcome of this step strongly depends on the accuracy of the demand forecast, and the potential for error is significant when designers have to guess the load in a new network. Therefore, many providers take a cautious approach by initially building a skeleton network and adding new resources as necessary by actual user demand. In this *incremental* network design, it is assumed that sets of user demands arrive over multiple time periods. Resources (e.g., OXCs, fiber links, wavelength channels) are added incrementally to satisfy each new set of demands, in a way that the additional capacity required is minimized.

A physical topology design problem was considered in [27]. Given a number of LSRs and a set of lightpaths to be set up among pairs of LSRs, the objective was to determine the two-connected physical topology with the minimum number of OXCs to establish all the lightpaths (this is a combined physical/virtual topology design problem in that the routing and wavelength assignment for the lightpaths is also determined). An iterative solution approach was considered, whereby a genetic algorithm was used to iterate over the space of physical topologies, and heuristics were employed for routing and wavelength assignment on a given physical topology (refer to the next subsection for details on RWA heuristics). The algorithm was applied to networks with up to 1000 LSRs and tens of thousands of lightpaths, and provided insight into the capacity requirements for realistic optical networks. For example, it was shown that the number of OXCs increases much slower than the number of LSRs, and also that the number of OXCs increases only moderately as the number of lightpaths increases by a factor of two or three. These results indicate that optical networks to interconnect a large number of LSRs can be built to provide rich connectivity with moderate cost.

Other studies related to capacity planning have looked into the problem of optimally placing network resources such as converters or power splitters (for multicast). The problem of converter placement was addressed in [11,28], and optimal [28] (for uniform traffic only) or near-optimal greedy [11] algorithms (for general traffic patterns) were developed. While both studies established that a small number of converters (approximately 30% of the number of OXCs) is sufficient, the results in [11] demonstrate that (a) the optimal placement of converters is extremely sensitive to the actual traffic pattern, and (b) an incremental approach to deploying converters may not lead to optimal (or near-optimal) results. The work in [29] considered the problem of optimally allocating the multicast-capable OXCs (MC-OXCs) to establish light-trees, and a greedy heuristic was proposed. It was found that there is little performance improvement if more than 50% of the OXCs in the network are multicast-capable, and that the optimal location of MC-OXCs depends on the traffic pattern.

Overall, the physical topology design problem is quite complex because the topology, the link and OXC capacities, and the number and location of optical devices such as converters and amplifiers strongly depends on the routing of lightpaths and the wavelength assignment strategy. If we make the problem less constrained, allowing the topology, routing, wavelength assignment, link capacity, etc., to change, the problem becomes very hard because these parameters are coupled in complicated ways. In practice, the topology may be constrained by external factors making the problem easier to deal with; for instance the existence of a deployed fiber infrastructure may dictate the location of OXCs and the links between them. However, the area of physical topology design for optical networks remains a rich area for future research.

Virtual Topology Design. A solution to the static RWA problem consists of a set of long-lived lightpaths which create a *logical* (or *virtual*) topology among the edge nodes. This virtual topology is embedded onto the physical topology of optical fiber links and OXCs. Accordingly, the static RWA problem is often referred to as the *virtual topology design* problem [9]. In the virtual topology, there is a directed link from edge node s to edge node d if a lightpath originating at s and terminating at d is set up (refer also to Figure 1), and edge node s is said to be "one hop away" from edge node d in the virtual topology, although the two nodes may be separated by a number of physical links. The type of virtual topology that can be created is usually constrained by the underlying physical topology. In particular, it is generally not possible to implement fully connected virtual topologies: for N edge nodes this would require each edge node to maintain $N-1$ lightpaths and the optical network to support a total of $N(N-1)$ lightpaths. Even for modest values of N, this degree of connectivity is beyond the reach of current optical technology, both in terms of the number of wavelengths that can be supported and in terms of the optical hardware (transmitters and receivers) required at each edge node.

In its most general form, the RWA problem is specified by providing the physical topology of the network and the traffic requirements. The physical topology corresponds to the deployment of cables in some existing fiber infrastructure, and is given as a graph $G_p(V, E_p)$, where V is the set of OXCs and E_p is the set of fibers that interconnect them. The traffic requirements are specified in a traffic matrix $\mathbf{T} = [\rho p_{sd}]$, where ρp_{sd} is a measure of the long-term traffic flowing from source edge node s to destination edge node d [30]. Quantity ρ represents the (deterministic) total offered load to the network, while the p_{sd} parameters define the distribution of the offered traffic.

Routing and wavelength assignment are considered together as an optimization problem using integer programming formulations. Usually, the objective of the formulation is to minimize the maximum congestion level in the network subject to network resource constraints [9,10]. While other objective functions are possible, such as minimizing the average weighted number of hops or minimizing the average packet delay, minimizing network congestion is preferable since it can lead to linear programming (ILP) formulations. While we do not present the RWA problem formulation here, the interested reader may refer to [30,9,10]. These formulations turn out to have extremely large numbers of variables, and are intractable for large networks. This fact has motivated the development of heuristic approaches for finding good solutions efficiently.

Before we describe the various heuristic approaches, we note that the static RWA problem can be logically decomposed into four subproblems. The decomposition is approximate or inexact, in the sense that solving the subproblems in sequence and combining the solutions may not result in the optimal solution for the fully integrated problem, or some later subproblem may have no solution given the solution obtained for an earlier subproblem, so no solution to the original problem may be obtained. However, the decomposition provides insight into the structure of the RWA problem and is a first step towards the design of effective heuristics. Assuming no wavelength conversion, the subproblems are as follows.

1. **Topology Subproblem:** Determine the logical topology to be imposed on the physical topology, that is, determine the lightpaths in terms of their source and destination edge nodes.
2. **Lightpath Routing Subproblem:** Determine the physical links which each lightpath consists of, that is, route the lightpaths over the physical topology.
3. **Wavelength Assignment Subproblem:** Determine the wavelength each lightpath uses, that is, assign a wavelength to each lightpath in the logical topology so that wavelength restrictions are obeyed for each physical link.
4. **Traffic Routing Subproblem:** Route packet traffic between source and destination edge nodes over the logical topology obtained.

A large number of heuristic algorithms have been developed in the literature to solve the general static RWA problem discussed here or its many variants. Overall, however, the different heuristics can be classified into three broad categories: (1) algorithms which solve the overall ILP problem sub-optimally, (2) algorithms which tackle only a subset of the four subproblems, and (3) algorithms which address the problem of embedding regular logical topologies onto the physical topology.

Suboptimal solutions can be obtained by applying classical tools developed for complex optimization problems directly to the ILP problem. One technique is to use LP-relaxation followed by rounding [31]. In this case, the integer constraints are relaxed creating a non-integral problem which can be solved by some linear programming method, and then a rounding algorithm is applied to obtain a new solution which obeys the integer constraints. Alternatively, genetic algorithms or simulated annealing [32] can be applied to obtain locally optimal solutions. The main drawback of these approaches is that it is difficult to control the quality of the final solution for large networks: simulated annealing is computationally expensive and thus, it may not be possible to adequately explore the state space, while LP-relaxation may lead to solutions from which it is difficult to apply rounding algorithms.

Another class of algorithms tackles the RWA problem by initially solving the first three subproblems listed above; traffic routing is then performed by employing well-known routing algorithms on the logical topology. One approach for solving the three subproblems is to maximize the amount of traffic that is carried on one-hop lightpaths, i.e., traffic that is routed from source to destination edge node directly on a lightpath. A greedy approach taken in [33] is to create lightpaths between edge nodes in order of decreasing traffic demands as long as the wavelength continuity and distinct wavelength constraints are satisfied. This algorithm starts with a logical topology with no links (lightpaths) and sequentially adds lightpaths as long as doing so does not violate any of the problem constraints. The reverse approach is also possible [34]: starting with a fully connected logical topology, an algorithm sequentially removes the lightpath carrying the smallest traffic flows until no constraint is violated. At each step (i.e., after removing a lightpath), the traffic routing subproblem is solved in order to find the lightpath with the smallest flow.

The third approach to RWA is to start with a given logical topology, thus avoiding to directly solve the first of the four subproblems listed above. Regular topologies are good candidates as logical topologies since they are well understood and results regarding bounds and averages (e.g., for hop lengths) are easier to derive. Algorithms for routing traffic on a regular topology are usually simple, so the traffic routing subproblem can be trivially solved. Also, regular topologies possess inherent load balancing characteristics which are important when the objective is to minimize the maximum congestion.

Once a regular topology is decided on as the one to implement the logical topology, it remains to decide which physical node will realize each given node in the regular topology (this is usually referred to as the *node mapping* subproblem), and which sequence of physical links will be used to realize each given edge (lightpath) in the regular topology (this *path mapping* subproblem is equivalent to the lightpath routing and wavelength assignment subproblems discussed earlier). This procedure is usually referred to embedding a regular topology in the physical topology. Both the node and path mapping subproblems are intractable, and heuristics have been proposed in the literature [34,35]. For instance, a heuristic for mapping the nodes of shuffle topologies based on the gradient algorithm was developed in [35].

Given that all the algorithms for the RWA problem are based on heuristics, it is important to be able to characterize the quality of the solutions obtained. To this end, one must resort to comparing the solutions to known bounds on the optimal solution. A comprehensive discussion of bounds for the RWA problem and the theoretical considerations involved in deriving them can be found in [9]. A simulation-based comparison of the relative performance of the three classes of heuristic for the RWA problem is presented in [10]. The results indicate that the second class of algorithms discussed earlier achieve the best performance.

The study in [22] also focused on virtual topology design (i.e., static RWA) for point-to-point traffic but observed that, since a light-tree is a more general representation of a lightpath, the set of virtual topologies that can be implemented using light-trees is a superset of the virtual topologies that can be implemented only using lightpaths. Thus, for any given virtual topology problem, an optimal solution using light-trees is guaranteed to be at least as good and possibly an improvement over the optimal solution obtained using only lightpaths. Furthermore, it was demonstrated that by extending the lightpath concept to a light-tree, the network performance (in terms of average packet hops) can be improved while the network cost (in terms of the number of optical transmitters/receivers required) decreases.

The static MC-RWA problem has been studied in [36,37]. The study in [36] focused on demonstrating the benefits of multicasting in wavelength routed optical networks. Specifically, it was shown that using light-trees (spanning the source and destination nodes) rather than individual parallel lightpaths (each connecting the source to an individual destination) requires fewer wavelengths and consumes a significantly lower amount of bandwidth. In [37] an ILP formulation that maximizes the total number of multicast connections was presented for the static MC-RWA problem. Rather than providing heuristic algorithms for solving the ILP, bounds on the objective function were presented by relaxing the integer constraints.

3.2 Dynamic RWA

During real-time network operation, edge nodes submit to the network requests for lightpaths to be set up as needed. Thus, connection requests are initiated in some random fashion. Depending on the state of the network at the time of a request, the available resources may or may not be sufficient to establish a lightpath between the corresponding source-destination edge node pair. The network state consists of the physical path (route) and wavelength assignment for all active lightpaths. The state evolves randomly in time as new lightpaths are admitted and existing lightpaths are released. Thus, each time a request is made, an algorithm must be executed in real time to determine whether it is feasible to accommodate the request, and, if so, to perform routing and wavelength assignment. If a request for a lightpath cannot be accepted because of lack of resources, it is blocked.

Because of the real-time nature of the problem, RWA algorithms in a dynamic traffic environment must be very simple. Since combined routing and wavelength assignment is a hard problem, a typical approach to designing efficient algorithms is to decouple the problem into two separate subproblems: the routing problem and the wavelength assignment problem. Consequently, most dynamic RWA algorithms for wavelength routed networks consist of the following general steps:

1. Compute a number of candidate physical paths for each source-destination edge node pair and arrange them in a path list.
2. Order all wavelengths in a wavelength list.
3. Starting with the path and wavelength at the top of the corresponding list, search for a feasible path and wavelength for the requested lightpath.

The specific nature of a dynamic RWA algorithm is determined by the number of candidate paths and how they are computed, the order in which paths and wavelengths are listed, and the order in which the path and wavelength lists are accessed.

Route Computation. Let us first discuss the routing subproblem. If a *static* algorithm is used, the paths are computed and ordered independently of the network state. With an *adaptive* algorithm, on the other hand, the paths computed and their order may vary according to the current state of the network. A static algorithm is executed off-line and the computed paths are stored for later use, resulting in low latency during lightpath establishment. Adaptive algorithms are executed at the time a lightpath request arrives and require network nodes to exchange information regarding the network state. Lightpath set up delay may also increase, but in general, adaptive algorithms improve network performance.

The number of path choices for establishing an optical connection is another important parameter. A *fixed* routing algorithm is a static algorithm in which every source-destination edge node pair is assigned a single path. With this scheme, a connection is blocked if there is no wavelength available on the designated path at the time of the request. In *fixed-alternate* routing, a number $k, k > 1$, of paths are computed and ordered off-line for each source-destination edge node pair. When a request arrives, these paths are examined in the specified order and the first one with a free wavelength is used to establish the lightpath. The request is blocked if no wavelength is available in any of the k paths. Similarly, an adaptive routing algorithm may compute a single path, or a number of alternate paths at the time of the request. A hybrid approach is to compute k paths off-line, however, the order in which the paths are considered is determined according to the network state at the time the connection request is made (e.g., least to most congested).

In most practical cases, the candidate paths for a request are considered in increasing order of *path length* (or *path cost*). Path length is typically defined as the sum of the weights (costs) assigned to each physical link along the path, and the weights are chosen according to some desirable routing criterion. Since weights can be assigned arbitrarily, they offer a wide range of possibilities for selecting path priorities. For example, in a static (fixed-alternate) routing algorithm, the weight of each link could be set to 1, or to the physical distance of the link. In the former case, the path list consists of the k minimum-hop paths, while in the latter the candidate paths are the k minimum-distance paths (where distance is defined as the geographic length). In an adaptive routing algorithm, link weights may reflect the load or "interference" on a link (i.e., the number of active lightpaths sharing the link). By assigning small weights to least loaded links, paths with larger number of free channels on their links rise to the head of the path list, resulting in a *least loaded* routing algorithm. Paths that are congested become "longer" and are moved further down the list; this tends to avoid heavily loaded bottleneck links. Many other weighting functions are possible.

When path lengths are sums of of link weights, the k-shortest path algorithm [38] can be used to compute candidate paths. Each path is checked in order of increasing length, and the first that is feasible is assigned the first free wavelength in the wavelength list. However, the k shortest paths constructed by this algorithm usually share links. Therefore, if one path in the list is not feasible, it is likely that other paths in the list with which it shares a link will also be infeasible. To reduce the risk of blocking, the k shortest paths can be computed so as to be pairwise link-disjoint. This can be accomplished as follows: when computing the i-th shortest path, $i = 1, \cdots, k$, the links used by the first $i - 1$ paths are removed from the original network topology and Dijkstra's shortest path algorithm [39] is applied to the resulting topology. This approach increases the chances of finding a feasible path for a connection request.

The problem of determining algorithms for routing multicast optical connections has also been studied in [37,40]. The problem of constructing trees for routing multicast connections was considered in [40] independently of wavelength assignment, under the assumption that not all OXCs are multicast capable, i.e., that there is a limited number of MC-OXCs in the network. Four new algorithms were developed for routing multicast connections under this *sparse light splitting* scenario. While the algorithms differ slightly from each other, the main idea to accommodate sparse splitting is to start with the assumption that all OXCs in the network are multicast capable and use an existing algorithm to build an initial tree. Such a tree is infeasible if a non-multicast-capable OXC is a branching point. In this case, all but one branches out of this OXC are removed, and destination nodes in the removed branches have to join the tree at a MC-OXC. In [37], on the other hand, the MC-RWA problem was solved by decoupling the routing and wavelength assignment problems. A number of *alternate* trees are constructed for each multicast connection using existing routing algorithms. When a request for a connection arrives, the associated trees are considered in a fixed order. For each tree, wavelengths are also considered in a fixed order (i.e., the first-fit strategy discussed in the next subsection). The connection is blocked if no free wavelength is found in any of the trees associated with the multicast connection.

We note that most of the literature (and the preceding discussion) has focused on the problem of obtaining paths that are optimal with respect to total path cost. In transparent optical networks, however, optical signals may suffer from physical layer impairments including attenuation, chromatic dispersion, polarization mode dispersion (PMD), amplifier spontaneous emission (ASE), cross-talk, and various nonlinearities [41]. These impairments must be taken into account when choosing a physical path. In general, the effect of physical layer impairments may be translated into a set of constraints that the physical path must satisfy; for instance, the total signal attenuation along the physical path must be within a certain power budget to guarantee a minimum level of signal quality at the receiver. Therefore, a simple shortest path first (SPF) algorithm (e.g., Dijkstra's algorithm implemented by protocols such as OSPF [42]) may not be appropriate for computing physical paths within a transparent optical network. Rather, constraint-based routing techniques such as the one employed by the constraint-based shortest path first (CSPF) algorithm [5] are needed. These techniques compute paths by taking into account not only the link cost but also a set of constraints that the path must satisfy. A first step towards the design of constraint-based routing algorithms for optical networks has been taken in [41] where it was shown how to translate the PMD and ASE impairments into a set of linear constraints on the end-to-end physical path. However, additional work is required to advance our understanding of how routing is affected by physical layer considerations, and constraint-based routing remains an open research area [43].

Wavelength Assignment. Let us now discuss the wavelength assignment subproblem which is concerned with the manner in which the wavelength list is ordered. For a given candidate path, wavelengths are considered in the order in which they appear in the list to find a free wavelength for the connection request. Again, we distinguish between the static and adaptive cases. In the static case, the wavelength ordering is fixed (e.g., the list is ordered by wavelength number). The idea behind this scheme, also referred to as *first-fit*, is to pack all the in-use wavelengths towards the top of the list so that wavelengths towards the end of the list will have higher probability of being available over long continuous paths. In the adaptive case, the ordering of wavelengths is typically based on usage. Usage can be defined either as the number of links in the network in which a wavelength is currently used, or as the number of active connections using a wavelength. Under the *most used* method, the most used wavelengths are considered first (i.e., wavelength are considered in order of decreasing usage). The rationale behind this method is to reuse active wavelengths as much as possible before trying others, packing connections into fewer wavelengths and conserving the spare capacity of less-used wavelengths. This in turn makes it more likely to find wavelengths that satisfy the continuity requirement over long paths. Under the *least used* method, wavelengths are tried in the order of increasing usage. This scheme attempts to balance the load as equally as possible among all the available wavelengths. However, least used assignment tends to "fragment" the availability of wavelengths, making it less likely that the same wavelength is available throughout the network for connections that traverse longer paths.

The most used and least used schemes introduce communication overhead because they require global network information in order to compute the usage of each wavelength. The first-fit scheme, on the other hand, requires no global information, and since it does not need to order wavelengths in real-time, it has significantly lower computational requirements than either the most used or least used schemes. Another adaptive scheme that avoids the communication and computational overhead of most used and least used is *random* wavelength assignment. With this scheme, the set of wavelengths that are free on a particular path is first determined. Among the available wavelengths, one is chosen randomly (usually with uniform probability) and assigned to the requested lightpath.

We note that in networks in which all OXCs are capable of wavelength conversion, the wavelength assignment problem is trivial: since a lightpath can be established as long as at least one wavelength is free at each link and different wavelengths can be used in different links, the order in which wavelengths are assigned is not important. On the other hand, when only a fraction of the OXCs employ converters (i.e., a sparse conversion scenario), a wavelength assignment scheme is again required to select a wavelength for each segment of a connection's path that originates and terminates at an OXC with converters. In this case, the same assignment policies discussed above for selecting a wavelength for the end-to-end path can also be used to select a wavelength for each path segment between OXCs with converters.

Performance of Dynamic RWA Algorithms. The performance of a dynamic RWA algorithm is generally measured in terms of the call blocking probability, that is, the probability that a lightpath cannot be established in the network due to lack of resources (e.g., link capacity or free wavelengths). Even in the case of simple network topologies (such as rings) or simple routing rules (such as fixed routing), the calculation of blocking probabilities in WDM networks is extremely difficult. In networks with arbitrary mesh topologies, and/or when using alternate or adaptive routing algorithms, the problem is even more complex. These complications arise from both the link load dependencies (due to interfering lightpaths) and the dependencies among the sets of active wavelengths in adjacent links (due to the wavelength continuity constraint). Nevertheless, the problem of computing blocking probabilities in wavelength routed networks has been extensively studied in the literature, and approximate analytical techniques which capture the effects of link load and wavelength dependencies have been developed in [11,19,16]. A detailed comparison of the performance of various wavelength assignment schemes in terms of call blocking probability can be found in [44].

Though important, average blocking probability (computed over all connection requests) does not always capture the full effect of a particular dynamic RWA algorithm on other aspects of network behavior, in particular, *fairness*. In this context, fairness refers to the variability in blocking probability experienced by lightpath requests between the various edge node pairs, such that lower variability is associated with a higher degree of fairness. In general, any network has the property that longer paths are likely to experience higher blocking than shorter ones. Consequently, the degree of fairness can be quantified by defining the *unfairness factor* as the ratio of the blocking probability on the longest path to that on the shortest path for a given RWA algorithm. Depending on the network topology and the RWA algorithm, this property may have a cascading effect which can result in an unfair treatment of the connections between more distant edge node pairs: blocking of long lightpaths leaves more resources available for short lightpaths, so that the connections established in the network tend to be short ones. These shorter connections "fragment" the availability of wavelengths, and thus, the problem of unfairness is more pronounced in networks without converters, since finding long paths that satisfy the wavelength continuity constraint is more difficult than without this constraint.

Several studies [11,19,16] have examined the influence of various parameters on blocking probability and fairness, and some of the general conclusions include the following:

- Wavelength conversion significantly affects fairness. Networks employing converters at all OXCs sometimes exhibit orders of magnitude improvement in fairness (as reflected by the unfairness factor) compared to networks with no conversion capability, despite the fact that the improvement in overall blocking probability is significantly less pronounced. It has also been shown that equipping a relatively small fraction (typically, 20-30%) of all OXCs with converters is sufficient to achieve most of the fairness benefits due to wavelength conversion.
- Alternate routing can significantly improve the network performance in terms of both overall blocking probability and fairness. In fact, having as few as three alternate paths for each connection may in some cases (depending on the network topology) achieve almost all the benefits (in terms of blocking and fairness) of having full wavelength conversion at each OXC with fixed routing.
- Wavelength assignment policies also play an important role, especially in terms of fairness. The random and least used schemes tend to "fragment" the wavelength availability, resulting in large unfairness factors (with least used having the worst performance). On the other hand, the most used assignment policy achieves the best performance in terms of fairness. The first-fit scheme exhibits a behavior very similar to most used in terms of both fairness and overall blocking probability, and has the additional advantage of being easier and less expensive to implement.

4   Control Plane Issues and Standardization Activities

So far we have focused on the application of network design and traffic engineering principles to the control of traffic in optical networks with a view to achieving specific performance objectives, including efficient utilization of network resources and planning of network capacity. Equally important to an operational network are associated control plane issues involved in automating the process of lightpath establishment and in supporting the network design and

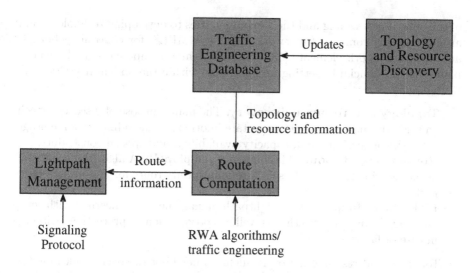

Fig. 5. Control plane components

traffic engineering functions. Currently, a number of standardization activities addressing the control plane aspects of optical networks are underway [45,46, 47] within the Internet Engineering Task Force (IETF) [48], the Optical Domain Service Interconnection (ODSI) coalition [49], and the Optical Internetworking Forum (OIF) [50]. In this section we review the relevant standards activities and discuss how they fit within the traffic engineering framework; we note, however, that these are ongoing efforts and will likely evolve as the underlying technology matures and our collective understanding of optical networks advances.

Let us return to Figure 1 which illustrates the manner in which client subnetworks (IP/MPLS networks in the figure) attach to the optical network of OXCs. The figure corresponds to the vision of a future optical network which is capable of providing a bandwidth-on-demand service by dynamically creating and tearing down lightpaths between client subnetworks. There are two broad issues that need to be addressed before such a vision is realized. First, a signaling mechanism is required at the user-network interface (UNI) between the client subnetworks and the optical network control plane. The signaling channel allows edge nodes to dynamically request bandwidth from the optical network, and supports important functions including service discovery and provisioning capabilities, neighbor discovery and reachability information, address registration, etc. Both the ODSI coalition [51] and the OIF [52] have developed specifications for the UNI; the OIF specifications are based on GMPLS [6].

Second, a set of signaling and control protocols must be defined within the optical network to support dynamic lightpath establishment and traffic engineering functionality; these protocols are implemented at the control module of each OXC. Currently, most of the work on defining control plane protocols in the optical network takes place under the auspices of IETF, reflecting a convergence of the optical networking and the IP communities to developing technology built around a single common framework, namely, GMPLS, for controlling both IP and optical network elements [53]. There are three components of the control plane that are crucial to setting up lightpaths within the optical network (refer to Figure 5):

- **Topology and resource discovery.** The main purpose of discovery mechanisms is to disseminate network state information including resource usage, network connectivity, link capacity availability, and special constraints.
- **Route Computation.** This component employs RWA algorithms and traffic engineering functions to select an appropriate route for a requested lightpath.
- **Lightpath Management.** Lightpath management is concerned with setup and tear-down of lightpaths, as well as coordination of protection switching in case of failures.

Topology and resource discovery includes neighbor discovery, link monitoring, and state distribution. The link management protocol (LMP) [54] has been proposed to perform neighbor discovery and link monitoring. LMP is expected to run between neighboring OXC nodes and can be used to establish and maintain control channel connectivity, monitor and verify data link connectivity, and isolate link, fiber, or channel failures. Distribution of state information is typically carried out through link state routing protocols such as OSPF [42]. There are currently several efforts under way to extend OSPF to support GMPLS [55] and traffic engineering [56]. In particular, the link state information that these protocols carry must be augmented to include optical resource information including: wavelength availability and bandwidth, physical layer constraints (discussed in Section 3.2), and link protection information, among others. This information is then used to build and update the optical network traffic engineering database (see Figure 5) which guides the route selection algorithm.

Once a lightpath is selected, a signaling protocol must be invoked to set up and manage the connection. Two protocols have currently been defined to signal a lightpath setup: RSVP-TE [57] and CR-LDP [58]. RSVP-TE is based on the resource reservation protocol (RSVP) [59] with appropriate extensions to support traffic engineering, while CR-LDP is an extension of the label distribution protocol (LDP) [60] augmented to handle constraint-based routing. The protocols are currently being extended to support GMPLS [61,62]. Besides signaling the path at connection time, both protocols can be used to automatically handle the switchover to the protection path once a failure in the working path has occurred. In the next section, we describe in detail the operation of some of these control plane protocols.

We note that the control plane elements depicted in Figure 5 are independent of each other and, thus, separable. This modularity allows each component to evolve independently of others, or to be replaced with a new and improved protocol. As the optical networking and IP communities come together to define standards, the constraints and new realities (e.g., the explosion in the number of channels in the network) imposed by the optical layer and WDM technology

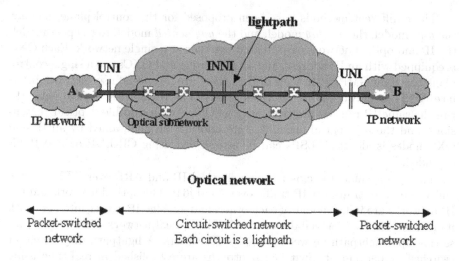

Fig. 6. IP networks interconnected by an optical network

will certainly affect our long-held assumptions regarding issues such as routing, control, discovery, etc., which have been developed for mostly opaque electronic networks. As we carefully rethink these issues in the context of transparent (or almost-transparent) optical networks, protocol design will certainly evolve to better accommodate the new technology. Therefore, we expect that the control plane protocols will continue to be refined and/or replaced by new, more appropriate ones. The interested reader should frequently check with the activities within IETF and OIF for the most recent developments.

We now proceed to describe some of the proposed control plane protocols. We first describe a proposed framework for transporting IP traffic over optical networks, and subsequently we present MPLS, LDP, CR-LDP, and GMPLS.

5   IP over Optical – A Framework

The issue of how IP traffic will be transported over an optical network has been addressed by IETF's IP over optical (IPO) working group in the [63]. The main features of this internet draft are summarized in this section.

An optical network is assumed to consist of interconnected optical sub-networks, where an optical sub-network consists of OXCs built by the same vendor. An optical network is a single administrative network. Optical networks can be combined to form an optical internetwork. An optical network is used as a backbone to interconnect a number of different IP networks, as well as other packet networks such as ATM and frame relay. In Figure 6, we show two IP networks interconnected via an optical network. We note that the edge IP router A is connected to the edge IP router B via a lightpath.

Three different methods have been proposed for the control plane, namely, the *peer* model, the *overlay* model, and the *augmented* model. In the peer model, the IP and optical networks are treated together as a single network. Each OXC is equipped with an IP address, and all IP routers and OXCs use a single control plane based on GMPLS. In view of this, there are no special user-network interface (UNI) or network-node interface (NNI). The IP and the optical networks run the same IP routing protocol, such as OSPF with suitable "optical" extensions, and the topological and link state information maintained by all IP and OXC nodes is identical. LSPs can be established using CR-LDP or RSVP-TE extended.

The overlay model is closer to the classical IP and ARP over ATM scheme which is used to transport IP traffic over ATM [64]. The optical network and the IP networks are independent of each other, and an edge IP router interacts with its ingress OXC over a well-defined UNI. The optical network is responsible for setting up a lightpath between two edge IP routers. A lightpath may be either switched or permanent. Switched lightpaths are established in real-time using signaling procedures, and they may last for a short or a long period of time. Permanent lightpaths are setup administratively by subscription and typically they last for a very long time. An edge IP router requests a switched lightpath from its ingress OXC using a signaling protocol over the UNI. Signaling messages are provided for creating, deleting, and modifying a switched lightpath. Routing within the optical network is independent of the routing within the IP networks.

Finally, in the augmented model the IP and optical networks use separate routing protocols, but information from one routing protocol is passed through the other routing protocol. For instance, external IP addresses could be carried within the optical routing protocol to allow reachability information to be passed to IP clients. The inter-domain IP routing protocol BGP may be used for exchanging information between IP and optical domains. Addressing of an OXC is identified by a unique IP address and a selector. The selector identifies further fine-grained information of relevance at the OXC, such as port, channel, sub-channel, etc. Typically, the setting up of a lightpath will be done in a distributed fashion similar to setting up a connection in ATM networks and also in MPLS-ready IP networks. Recently, it has been proposed to use a centralized scheme for setting up lightpaths. This requires a centralized server which has complete knowledge of the physical topology and wavelength availability [65,66]. The Common Open Policy Service (COPS) signaling protocol is used by the ingress switch of an edge router to request the establishment of a connection from the Policy Decision Point (PDP), a remote server, which calculates the path and downloads the information to all the nodes along the path.

5.1   Multiprotocol Label Switching (MPLS)

In order to understand the signalling protocols that have been proposed to control a wavelength-routed optical network, we first need to examine the Multiprotocol Label Switching (MPLS) scheme.

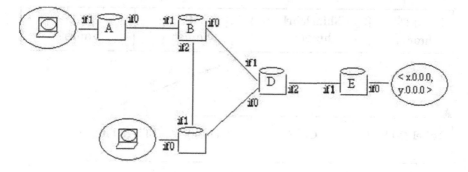

The shim label header

MPLS was developed as a means of introducing connection oriented features in an IP network. A router forwards an IP packet according to its prefix. In a given router, the set of all addresses that have the same prefix, is referred to as the forwarding equivalent class (FEC). IP packets belonging to the same FEC have the same output interface. In MPLS, a FEC is associated with a label. This label is used to determine the output interface of an IP packet without having to do the traditional look-up its address in the routing table. In IPv6, the label can be carried in the flow label field. In IPv4, however, there is no space for such a label in the IP header. If the IP network runs on top of an ATM network, the label is carried in the VPI/VCI field of an ATM cell. If it is running over frame relay, the label is carried in the DLCI field. For Ethernet, token ring, and point-to-point connections running a link layer protocol such as PPP, the label is carried in a special shim label header, which is inserted between the LLC header and the IP header, as shown in Figure 7. The first field of the shim label header is a 20-bit field used to carry the label. The second field is a 3-bit field used for the class-of-service (CoS) indication. This field is used to indicate the priority of the IP packet. The S field is used in conjunction with the label stack. Finally, the time-to-live (TTL) field is similar to the TTL field in the IP header. A label switching network consists of label switching routers (LSR), which are IP routers that run MPLS, they forward IP packets based on their labels, and they can also carry the customary IP forwarding decision based on the prefix of an IP addresses. An MPLS node is an LSR which may not necessarily forward IP packets based on the prefixes.

To see how label switching works, let us consider a network consisting of 5 LSRs, A, B, C, D, and E, linked with point-to-point connections as shown in Figure 8. We assume that a new set of hosts with the prefix <x.0.0.0,y.0.0.0>, where x.0.0.0 is the base network address and y.0.0.0 is the mask, is directly connected to E. The flow of IP packets with this prefix from A to E is via B and D. That is, A's next-hop router for this prefix is B, B's next-hop router is D, and D's next-hop router is E. Likewise, the flow of IP packets with the same prefix from C to E is via D. That is, C's next-hop router for this prefix is D, and D's next-hop router is E. The interfaces in Figure 8 show how these routers

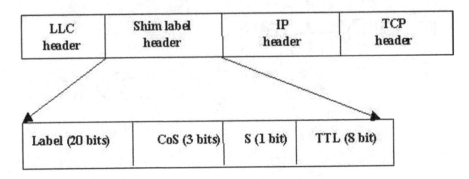

Fig. 8. An example of label switching

are interconnected. For instance, A is connected to B via if0, B is connected to A via if1, to C via if2 and to D via if0, and so on. When an LSR identifies the FEC associated with this new prefix <x.0.0.0,y.0.0.0>, it selects a label from a pool of free labels and it makes an entry into a table known as the *label forward information base* (LFIB). This table contains information regarding the incoming and outgoing labels associated with a FEC, the output interface, i.e., the FEC's next-hop router, and the operation that needs to be performed on the label stack. Incoming IP packets belonging to this particular FEC have to be labeled with the value selected by the LSR. In view of this, the LSR has to notify its neighbours about its label selection for the particular FEC. In the above example, LSR B sends its information to A, D, and C. A recognizes that it is upstream from B, and it uses the information to update the entry for this FEC in its LFIB. D sends its information to B, C, and E. Since B and C are both upstream of D, they use this information to update the entries in their LFIB. E sends its information to D, which uses it to update its entry in its LFIB. As a result, in each LSR each incoming label associated with a FEC is bound to an outgoing label in the LFIB entry. In Figure 9, we show the labels allocated by the LSRs. The sequence of labels 62, 15, 60 forms a path, referred to as the label switched path (LSP). Typically, there may be several label switched paths associated with the same FEC which form a tree, as shown in Figure 9.

Once the labels have been distributed and the entries have been updated in the LFIBs, the forwarding of an IP packet belonging to the FEC associated with the prefix <x.0.0.0, y.0.0.0> is done using solely the labels. Let us assume that A receives an IP packet from one of its local hosts with a prefix <x.0.0.0, y.0.0.0>. A identifies that the packet's IP address belongs to the FEC, and it looks up its LFIB to obtain the label value and the outgoing interface. It creates a shim label header, sets the label value to 62, and forwards it to the outgoing interface if0. When the IP packet arrives at LSR B, its label is extracted and looked up in B's LFIB. The old label is replaced by the new one, which is 15, and the IP packet is forwarded to interface if0. LSR D follows exactly the same procedure. When it receives the IP packet from B, it replaces its incoming label with the outgoing

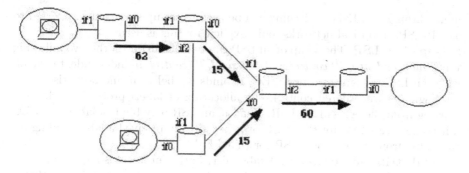

Fig. 9. Label switched paths

label, which is 60, and forwards it to interface if2. Finally, E forwards the IP packet to its local destination. The same procedure applies for an IP packet with a prefix <x.0.0.0, y.0.0.0> that arrives at C. Labeled IP packets within an LSR are served according to their priority, carried in the CoS field of he shim header. Specifically, an IP router maintains different quality-of-service queues for each output interface. These queues are served using a scheduling algorithm, so that different classes of IP packets can be served according to their requested quality of service. Another interesting feature of label switching is that it can be used to create a dedicated route, known as an explicit route, between two IP routers. Explicit routing is used primarily in optical networks, and it is described below.

Label allocation. In the example described above, a label is generated by the LSR which is at the downstream end of the link, with respect to the flow of the IP packets. In view of this, this type of label allocation is known as *downstream label allocation*. In addition to this scheme, labels can be allocated using *downstream label allocation on demand*. In this case, each LSR allocates an incoming label to a FEC and creates an appropriate entry in its LFIB. However, it does not advertise its label to its neighbours as in the case of downstream allocation. Instead, an upstream LSR obtains the label information by issuing a request.

Explicit routing. As we have discussed above, a router makes a forwarding decision by using the IP address in its routing table in order to determine the next-hop router. Typically, each IP router calculates the next-hop router for a particular destination using the shortest path algorithm. Label switching follows the same general approach, only it uses labels. This routing scheme is known as hop-by-hop routing. An alternative way of routing a packet is to use source routing. In this case, the originating (source) LSR selects the path to the destination LSR. Other LSRs on the path simply obey the source's routing instructions. Source routing can be used in an IP network for a variety of reasons, such as to evenly distribute traffic among links by moving some of the traffic from highly utilized links to less utilized links (load balancing), create tunnels for MPLS-based VPNs, and introduce routes based on a quality-of-service criterion such as minimize the number of hops, minimize the total end-to-end delay, and maximize throughput. Label switching can be used to set-up such routes, referred to as CR-LSP. In optical networks, only explicit routing is used.

Set-up of an LSP. The setup of an LSP can be done in one of the two following ways: *independent LSP control* and *ordered LSP control*. In independent control, when an LSR recognizes a new FEC, it binds a label to it and advertises it to its neighbors. In ordered control, the allocation of labels proceeds backwards starting from the egress LSP LSR. That is, an LSR only binds a label to a FEC if it is the egress LSR for that FEC or it has already received a label binding for that FEC from its next hop LSR for that FEC.

Label distribution protocol. A label distribution protocol is required to reliably establish and maintain label bindings. As mentioned previously, the RSVP protocol and its extension RSVP-TE have been proposed to be used as label distribution protocol. In addition, we new protocol, the label distribution protocol (LDP), has been proposed for MPLS. LDP has been extended to CR-LDP for the establishment, maintenance, and tearing down of explicit routes.

5.2   The Label Distribution Protocol (LDP)

For reliability purposes, the LDP protocol runs over TCP [60]. Two LSRs that run LDP and they are directly connected are known as LDP peers.

An LSR discovers potential LDP peers by sending periodically LDP link hello messages out of each interface. The receipt of an LDP hello message triggers the establishment of an LDP session between two LDP peers. When the LDP session is initialized, the two LDP peers negotiate session parameters such as label distribution method, timer values, range of VPI/VCI values for ATM, and range of DLCI values for frame relay. An LDP session is soft-state and it needs to be continuously refreshed. A session is maintained as long as traffic flows (in the form of LDP PDUs) over the session. In the absence of LDP PDUs, keepAlive messages are sent. LDP supports independent label distribution control and ordered label distribution control. It also provides functionality for detection of loops in the LSP. Information in LDP is sent in the form of LDP PDUs, which consists of a header followed by one or more LDP messages. An LDP message consists of a header followed by mandatory and optional parameters. The header and the parameters are all encoded using the type-length-value (TLV) scheme. The type specifies how the value field is to be interpreted, the length gives the length of the value, and the value field contains the actual information. The value field contains one or more TLVs.

The following LDP messages have been defined: notification, hello, initialization, keepAlive, address, address withdraw, label mapping, label request, label abort, label withdraw, and label release. The notification message is used to inform an LDP peer of a fatal error or to provide advisory information regarding the outcome of processing an LDP message. The hello messages are used to discover peer LDPs, and the initialization message is used to initialize a new session between two peer LDPs. The address message is used by an LSR to advertise the address of its interfaces. Previously advertised addresses can be withdrawn using the address withdraw message. The label mapping message is used by an LSR to advertise a binding of a label to a FEC to its LDP peers. The label request message is used by an LSR to request a label from a peer LDP to a FEC. An LSR may transmit a request message under the following conditions:

- The LSR recognizes a new FEC via the forwarding table, and the next hop is an LDP peer, and the LSR does not already have a mapping from the next hop for the given FEC.
- The next hop to the FEC changes, and the LSR does not already have a mapping from the next hop for the given FEC.
- The LSR receives a label request for a FEC from an upstream LDP peer, the FEC next hop is an LDP peer, and the LSR does not already have a mapping from the next hop.

5.3 Constrained Routing Label Distribution Protocol (CR-LDP)

CR-LDP is a signaling protocol based on LDP, and it runs over TCP. It is used to set-up a point-to-point LSP, referred to as CR-LSP. A CR-LSP, unlike an LSP, is a point-to-point path through an MPLS network, which is set-up based on criteria not limited to routing information, such as explicit routing and QoS based routing. A CR-LDP may be used for a variety of reasons, such as, to evenly distribute traffic among links (load balancing), create tunnels for MPLS-based VPNs, and introduce routes based on a QoS criteria, such as minimization of the number of hops, minimization of the total end-to-end delay, and maximization of throughput.

A CR-LSP is setup as follows. A request at an ingress LSR to setup a CR-LSP originates from a management system or an application. The ingress LSR calculates the explicit route using information provided by the management system, or the application, or form a routing table. The explicit route is a series of nodes or groups of nodes (referred to as abstract nodes), which is signalled to nodes or abstract nodes along the path using the label request message. CR-LSPs are set up using ordered control with downstream on demand label allocation. Strict and loose explicit routes can be used. In a strict route all the LSRs through which the CR-LSP must pass are indicated. In loose routing, some LSRs are specified, and the exact path between two such LSRs is determined using conventional routing based on IP addresses. Route pinning is a feature that can be used to fix the path through a loosely defined route, so that it does not change when a better next hop becomes available.

As in ATM networks, CR-LDP permits the specification of traffic parameters for a CR-LSP. The following five traffic parameters have been specified: peak data rate (PDR), peak burst size (PBS), committed data rate (CDR), committed bucket size (CBS), and excessive bucket size (EBS). PBS and PDR are used to specify the peak rate. This is the maximum rate at which traffic is sent to the CR-LSP, and it is expressed in bytes/sec. It is defined in terms of a token bucket whose maximum bucket size is set equal to the peak burst size (PBS) and the rate at which it is replenished is equal to the peak data rate (PDR). CBS and CDR are used to specify the committed rate, which is the amount of bandwidth allocated to a CR-LSP by an LSR. It is defined by a token bucket whose maximum bucket size is set equal to CBS and the rate at which the bucket is replenished is equal to CDR. Finally, the excess bucket size (EBS) is used to define the maximum size of a third token bucket, the excess token bucket, which is replenished at the rate of CDR. By appropriately manipulating the values of these five traffic parameters, it is possible to establish different classes of service.

The establishment of an CR-LSP is achieved using the label request message and label mapping message. The label request message carries the list of all nodes and abstract nodes which are on the path of the CR-LSP, the traffic parameters, FEC, and other relevant parameters. It is propagated from the ingress LSR to the egress LSR. The label mapping message is used to advertise the labels, which is done using ordered control, that is from the egress LSR back towards the ingress LSR.

5.4   Generalized MPLS (GMPLS)

GMPLS [6] extends the label switching architecture proposed in MPLS to other types of non-packet based networks, such as SONET/SDH based networks and wavelength-routed networks. Specifically, the GMPLS architecture supports the following types of switching: packet switching (IP, ATM, and frame relay), wavelength switching in a wavelength-routed network, port or fiber switching in a wavelength-routed network, and time slot switching for a SONET/SDH cross-connect.

A GMPLS LSR may support the following five interfaces: packet switch interfaces, layer-2 switch interfaces, time-division multiplex interfaces, lambda switch interfaces, and fiber switch interfaces. A packet switch interface recognizes packet boundaries and it can forward packets based on the content of the IP header or the content of the shim header. A layer-2 switch interface recognizes frame/cell boundaries and can forward data based on the content of the frame/cell header. Examples include interfaces on ATM-LSRs that forward cells based on their VPI/VCI value, and interfaces on Ethernet bridges that forward data based on the MAC header. A time-division multiplex interface forwards data based on the data's time slot in a repeating cycle (frame). Examples of this interface is that of a SONET/SDH cross-connect, terminal multiplexer, and add-drop multiplexer. Other examples include interfaces implementing the digital wrapper (G.709) and PDH interfaces. A lamda-switch interface forwards the optical signal from an incoming wavelength to an outgoing wavelength. An example of such an interface is the optical cross-connect (OXC) that operates at the level of an individual wavelength or a group of wavelengths (waveband). Finally, a fiber switch interface forwards the signals from one (or more) incoming fibers to one (or more) outgoing fibers. An example of this interface is an OXC that operates at the level of a fiber or group of fibers.

GMPLS extends the control plane of MPLS to support each of the five classes of interfaces. The GMPLS supports the peer model, the overlay model and the augmented model. In GMPLS, downstream on-demand label allocation is used with ordered control initiated by an ingress node. There is no restriction on the route selection. Explicit routing is normally used, but hop-by-hop routing can be also used. There is also no restriction on the way an LSP is set-up. It could be set-up as described in the example above in the MPLS section (control driven), or it could be set-up as a result of a user issuing a request to establish an LSP. The latter approach is suitable for circuit-switching technologies. Several new forms of labels are required to deal with the widened scope of MPLS into the optical and time division multiplexing domain. The new label not only allows for the familiar label that travels in-band with the associated packet, but it also allows for labels which identify time-slots, wavelengths, or fibers. The generalized label may carry a label that represents a single fiber in a bundle, a single wavelength within a fiber, a single wavelength within a waveband or a fiber, a set of time-slots within a the SONET/SDH payload carried over a wavelength, and the MPLS labels for IP packets, ATM cells and frame relay frames. This new label is known as the generalized label.

CR-LDP [62] and RSVP-TE [61] have both been extended to allow the signalling and instantiation of lightpaths. A UNI signalling protocol has been proposed by OIF based on GMPLS. The interior gateway protocols IS-IS and OSPF have been extended to advertise availability of optical resources (i.e., bandwidth on wavelengths, interface types) and other network attributes and constraints. Also, a new link management protocol (LMP) has been developed to address issues related to the link management in optical networks.

6   Optical Packet Switching

Optical packet switching has been proposed as a solution to transporting packets over an optical network. Optical packet switching is sometimes referred to as "optical ATM," since it resembles ATM, but it takes place in the optical domain.

A WDM optical packet switch consists of four parts, namely, the input interface, the switching fabric, the output interface, and the control unit. The input interface is mainly used for packet delineation and alignment, packet header information extraction and packet header removal. The switch fabric is the core of the switch and it is used for switching packets optically. The output interface is used to regenerate the optical signals and insert the packet header. The control unit controls the switch using the information in the packet headers. Because of synchronization requirements, optical packet switches are typically designed for fixed-size packets.

When a packet arrives at a WDM optical packet switch, it is first processed by the input interface. The header and the payload of the packet are separated, and the header is converted into the electrical domain and processed by the control unit electronically. The payload remains as an optical signal throughout the switch. After the payload passes through the switching fabric, it is re-combined with the header, which has been converted back into the optical domain, at the output interface.

In the following, we briefly describe some issues of optical packet switches. For more information about synchronization and contention resolution, the reader is referred to [67].

Packet coding techniques. Several optical packet coding techniques have been studied. There are three basic categories, namely, bit-serial, bit-parallel, and out-of-band-signaling. Bit-serial coding can be implemented using optical code division multiplexing (OCDM), or optical pulse interval, or mixed rate techniques. In OCDM, each bit carries its routing information, while in the latter two techniques, multiple bits are organized into a packet payload with a packet header that includes routing information. The difference between the latter two techniques is that in optical pulse interval the packet header and payload are transmitted at the same rate, whereas in mixed rate technique the packet header is transmitted at a lower rate than the payload so that the packet header can be easily processed electronically. In bit-parallel coding, multi-bits are transmitted at the same time but on separate wavelengths. Out-of-band-signaling coding includes sub-carrier multiplexing (SCM) and dual wavelength coding. In SCM, the packet header is placed in an electrical subcarrier above the baseband frequencies occupied by the packet payload, and both are transmitted at the same time slot. In dual wavelength coding, the packet header and payload are transmitted in separate wavelengths but at the same time slot.

Contention resolution. Contention resolution is necessary in order to handle the case where more than one packet are destined to go out of the same output port at the same time. This is a problem that commonly arises in packet switches, and it is known as *external blocking* [68]. It is typically resolved by buffering all the contending packets, except one which is permitted to go out. In an optical packet switch, techniques designed to address the external blocking problem include *optical buffering, exploiting the wavelength domain*, and using *deflection routing*. Whether these prove to be adequate to address the external blocking problem is still highly doubtful. Below we discuss each of these solutions.

Optical buffering currently can only be implemented using optical delay lines (ODL). An ODL can delay a packet for a specified amount of time, which is related to the length of the delay line. Currently, optical buffering is the Achilles' heel of optical packet switches! Delay lines may be acceptable in prototype switches, but they are not commercially viable. The alternative, of course, is to convert the optical packet to the electrical domain and store it electronically. This is not an acceptable solution, since electronic memories cannot keep up with the speeds of optical networks.

There are many ways that an ODL can be used to emulate an electronic buffer. For instance, a buffer for N packets with a FIFO discipline can be implemented using N delay lines of different lengths. Delay line i delays a packet for i timeslots. A counter keeps track of the number of the packets in the buffer. It is decremented when a packet leaves the buffer, and it is incremented when a packet enters the buffer. Suppose that the value of the counter is j when a packet arrives at the buffer, then the packet will be routed to the j-th delay line.

Limited by the length of the delay lines, this type of buffer is usually small, and it does not scale up.

An alternative solution to optical buffering is to use the wavelength domain. In WDM, several wavelengths run on a fiber link that connects two optical switches. This can be exploited to minimize external blocking as follows. Let us assume that two packets are destined to go out of the same output port at the same time. Then, they can be still transmitted out but on two different wavelengths. This method may have some potential in minimizing external blocking, particularly since the number of wavelengths that can be coupled together onto a single fiber continues to increase. For instance, it is expected that in the near future there will be as many as 200 wavelengths per fiber. This method requires wavelength converters.

Finally, deflection routing is another alternative to solving the external blocking problem. Deflection routing is ideally suited to switches that have little buffer space. When there is a conflict between two packets, one will be routed to the correct output port, and the other will be routed to any other available output port. In this way, no or little buffer is needed. However, the deflected packet may end up following a longer path to its destination. As a result, the end-to-end delay for a packet may be unacceptably high. Also, packets will have to be re-ordered at the destination since they are likely to arrive in an out-of-sequence manner.

6.1   Optical Packet Switch Architectures

Various optical packet switch architectures that have been proposed in the literature. For a review of some of these architectures see [69] . Based on the switching fabric used, they have been classified in the following three classes: space switch fabrics, broadcast-and-select switch fabrics, and wavelength routing switch fabrics. For presentation purposes, we only give an example below based on a space switch fabric.

An architecture with a space switch fabric. A space switch fabric architecture is shown in Figure 10. The performance of this switch was analyzed in [70]. The switch consists of N incoming and N outgoing fiber links, with n wavelengths running on each fiber link. The switch is slotted, and the length of the slot is such that an optical packet can be transmitted and propagated from an input port to an output optical buffer.

The switch fabric consists of three parts: optical packet encoder, space switch, and optical packet buffer. The optical packet encoder works as follows. For each incoming fiber link, there is an optical demultiplexer which divides the incoming optical signal to the n different wavelengths. Each wavelength is fed to a different tunable wavelength converter (TWC) which converts the wavelength of the optical packet to a wavelength that is free at the destination optical output buffer. Then, through the space switch fabric, the optical packet can be switched to any of the N output optical buffers. Specifically, the output of a TWC is fed to a splitter which distributes the same signal to N different output fibers, one per output buffer. The signal on each of these output fibers goes through another

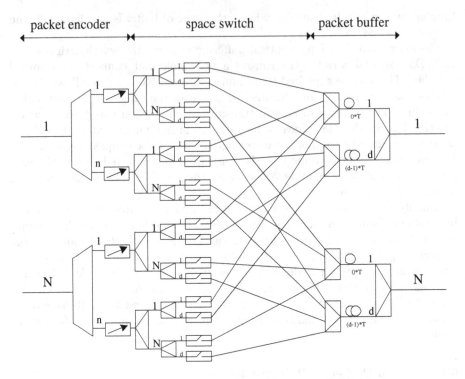

Fig. 10. An architecture with a space switch fabric

splitter which distributes it to $d + 1$ different output fibers, and each output fiber is connected through an optical gate to one of the ODLs of the destination output buffer. The optical packet is forwarded to an ODL by appropriately keeping one optical gate open, and closing the remaining. The information regarding which wavelength a TWC should convert the wavelength of an incoming packet and the decision as to which ODL of the destination output buffer the packet will be switched to is provided by the control unit, which has knowledge of the state of the entire switch.

Each output buffer is an optical buffer implemented as follows. It consists of $d + 1$ ODLs, numbered from 0 to d. ODL i delays an optical packet for a fixed delay equal to i slots. ODL 0 provides zero delay, and a packet arriving at this ODL is simply transmitted out of the output port. Each ODL can delay optical packets on each of the n wavelengths. For instance, at the beginning of a slot, ODL 1 can accept up to n optical packets, one per wavelength, and delay them for 1 slot. ODL 2 can accept up to n optical packets at the beginning of each time slot, and delay them for 2 slots. That is, at slot t, it can accept up to n packets (one per wavelength) and delay them for 2 slots, in which case, these packets will exit at the beginning of slot $t + 2$. However, at the beginning of slot $t + 1$, it can also accept another batch of n optical packets. Thus, a maximum of

$2n$ packets may be in transit within ODL 2. Similarly for ODL 3 through d. Let c_i denote the number of optical packets on wavelength i, where $i = 1, 2, \cdots, n$. We note that these c_i optical packets may be on a number of different ODLs. To insert an optical packet into the buffer, we first check all the c_i counters to find the smallest one, say c_s, then we set the TWC associated with this optical packet to convert the packet's wavelength to s, increase c_s by one, and switch the optical packet to ODL c_s. If the smallest counter c_s is larger than d, the packet will be dropped.

7   Optical Burst Switching

Optical burst switching (OBS) is a technique for transmitting bursts of traffic through an optical transport network by reserving resources through the optical network for only one burst. This technique is an adaptation of an ITU-T standard for burst switching for ATM networks, known as ATM block transfer (ABT) [64]. It is a new technology that has not as yet been commercialized. The main idea of OBS is shown in Figure 11. End-devices A and B communicate via a network of OBS nodes by transmitting data in bursts. An OBS node can be seen as consisting of a switch fabric and a CPU which controls the switch fabric and also processes signalling messages. The switch fabric is an $N \times N$ switch, where each incoming or outgoing fiber has W wavelengths, and it switches incoming bursts to their requested output ports. It may or may not be equipped with converters. Early proposals for OBS required an OBS node to be equipped with optical buffers. However, more recently it has been proposed to use bufferless OBS nodes.

Let us consider now the flow of bursts from end-device A to B. For each burst, A first sends a SETUP message via a signaling channel [71] to its ingress switch announcing its intention to transmit a burst. Transmission of the burst takes place after a delay known as offset. The ingress switch processes the SETUP message and allocates resources in its switch fabric so that to switch the burst out of the destination output port. The SETUP message is then forwarded to the next OBS node, which processes the SETUP message and allocates resources to switch the burst through its switch fabric. This goes one until the SETUP message reaches the destination end-device B. Each node in the path of the burst allocates resources to switch the burst through its switch for just a single burst, and it frees these resources after the burst has come through. A burst is dropped if an OBS node does not have enough capacity to switch it through its switch fabric.

The burst may last a short period of time and it may contain several packets, such as IP packets, ATM cells, and frame relay frames. It may also last for a long time, like a lightpath. In view of this, OBS can be seen as lying in-between packet switching and circuit switching.

Several variants of OBS have been proposed, such as tell-and-go (TAG), tell-and-wait (TAW), just-enough-time (JET) [72], and just-in-time (JIT) [73,71]. In the tell-and-go scheme, the source transmits the SETUP message and imme-

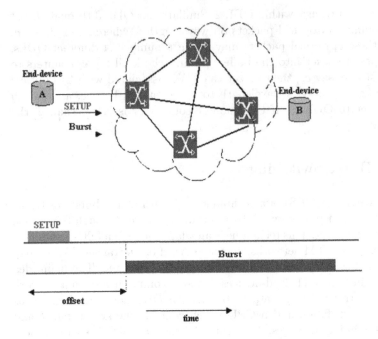

Fig. 11. An example of optical burst switching

diately after it transmits the optical burst. The tell-and-go scheme is inspired by one of the variants of the ATM block transfer (ABT) scheme in ATM network. However, within the setting of OBS, it is a rather idealistic scheme since there is no time for the receiving OBS node to process the SETUP message and to configure its switch fabric on time so that to transmit the incoming burst to its destination output port. In order to implement this scheme, either the OBS node has already been configured to switch the burst or some input optical buffering may be required to hold the burst until the node can process the SETUP message.

The tell-and-wait (TAW) scheme is the opposite of TAG, and it was inspired by another variant of the ATM block transfer (ABT) scheme. In this case, the SETUP message is propagated all the way to the receiving end-device, and each OBS node along the path processes the SETUP message and allocates resources within its switch fabric. A positive acknowledgement is returned to the transmitting end-device, upon receipt of which the end-device transmits its bursts. In this case, the burst will go through without been dropped at any OBS node. The offset can be seen as being equal to the round trip propagation delay plus the sum of the processing delays of the SETUP message at each OBS node along the path. In the just-enough-time (JET) and just-in-time (JIT) schemes, there is a delay between the transmission of the control packet and the transmission of the optical burst. This delay has to be larger than the sum of the

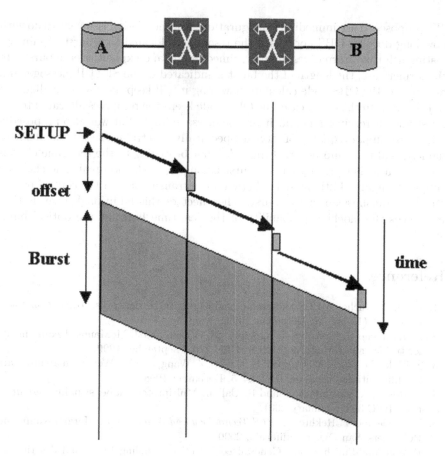

Fig. 12. The offset has to be larger than the sum of processing times

total processing times of the SETUP message at all the OBS nodes along the path. In this way, when the burst arrives at an OBS node, the SETUP message has already been processed and resources have been allocated so as to switch the burst through the switch fabric. An example is shown in Figure 12. The two schemes JET and JIT vary significantly in their proposed implementation. One of the main design issues has to do with the time at which an OBS node should configure its switch fabric to receive the pending burst. There are two alternatives, namely, *immediate configuration* and *estimated configuration*. In the former case, the OBS node allocates resources to the incoming burst immediately after it processes the SETUP message, whereas in the latter case, it allocates the necessary resources later on at a time when it estimates that the burst will arrive at the node. Obviously, in the case of immediate configuration, resources go unused until the burst arrives, whereas in the estimated configuration scheme the resources are better utilized. However, the immediate configuration scheme is considerably simpler to implement. JET uses estimated configuration, whereas

JIT proposes to use immediate configuration. Another design issue has to do with how long does an OBS node keep the resources allocated to a burst. Again, we distinguish two alternatives, namely, timed bursts and explicit release bursts. In the former case, the length of the burst is indicated in the SETUP message, and as a result, the OBS node calculate how long it will keep its resources allocated to the burst. In the latter case, the OBS node keeps the resources allocated to the burst until it receives an explicit release message. In JET it was also proposed a scheme to supports quality of service. Specifically, two traffic classes were defined, namely, real-time and non-real-time. A burst belonging to the real-time class is allocated a higher priority than a burst belonging to the non-real-time class, by simply using an additional delay between the transmission of the control packet and the transmission of the burst. The effect of this additional delay is that it reduces the blocking probability of the real-time burst at the optical burst switch.

References

1. D. O. Awduche. MPLS and traffic engineering in IP networks. *IEEE Communications*, 37(12):42–47, December 1999.
2. D. Awduche, J. Malcolm, J. Agogbua, M. O'Dell, and J. McManus. Requirements for traffic engineering over MPLS. RFC 2702, September 1999.
3. S. Blake, D. Black, M. Carlson, E. Davies, Z. Wang, and W. Weiss. An architecture for differentiated services. RFC 2475, December 1998.
4. E. Rosen, A. Viswanathan, and R. Callon. Multiprotocol label switching architecture. RFC 3031, January 2001.
5. B. Davie and Y. Rekhter. *MPLS Technology and Applications*. Morgan Kaufmann Publishers, San Diego, California, 2000.
6. P. Ashwood-Smith *et al.* Generalized MPLS – signaling functional description. IETF Draft <draft-ietf-mpls-generalized-signaling-06.txt>, April 2001. Work in progress.
7. D. H. Su and D. W. Griffith. Standards activities for MPLS over WDM networks. *Optical Networks*, 1(3), July 2000.
8. O. Gerstel, B. Li, A. McGuire, G. N. Rouskas, K. Sivalingam, and Z. Zhang (Eds.). Special issue on protocols and architectures for next generation optical WDM networks. *IEEE Journal Selected Areas in Communications*, 18(10), October 2000.
9. R. Dutta and G. N. Rouskas. A survey of virtual topology design algorithms for wavelength routed optical networks. *Optical Networks*, 1(1):73–89, January 2000.
10. E. Leonardi, M. Mellia, and M. A. Marsan. Algorithms for the logical topology design in WDM all-optical networks. *Optical Networks*, 1(1):35–46, January 2000.
11. Y. Zhu, G. N. Rouskas, and H. G. Perros. A path decomposition approach for computing blocking probabilities in wavelength routing networks. *IEEE/ACM Transactions on Networking*, 8(6):747–762, December 2000.
12. L. Li and A. K. Somani. A new analytical model for multifiber WDM networks. *IEEE Journal Selected Areas in Communications*, 18(10):2138–2145, October 2000.
13. S. Ramamurthy and B. Mukherjee. Survivable WDM mesh networks, part I – protection. In *Proceedings of INFOCOM '99*, pages 744–751, March 1999.
14. S. Ramamurthy and B. Mukherjee. Survivable WDM mesh networks, part II – restoration. In *Proceedings of ICC '99*, pages 2023–2030, June 1999.

15. A. Mokhtar and M. Azizoglu. Adaptive wavelength routing in all-optical netowrks. *IEEE/ACM Transactions on Networking*, 6(2):197–206, April 1998.

16. E. Karasan and E. Ayanoglu. Effects of wavelength routing and selection algorithms on wavelength conversion gain in WDM optical networks. *IEEE/ACM Transactions on Networking*, 6(2):186–196, April 1998.

17. H. Zang, J. P. Jue, and B. Mukherjee. A review of routing and wavelength assignment approaches for wavelength-routed optical WDM networks. *Optical Networks*, 1(1):47–60, January 2000.

18. B. Ramamurthy and B. Mukherjee. Wavelength conversion in WDM networking. *IEEE Journal Selected Areas in Communications*, 16(7):1061–1073, September 1998.

19. S. Subramaniam, M. Azizoglu, and A. Somani. All-optical networks with sparse wavelength conversion. *IEEE/ACM Transactions on Networking*, 4(4):544–557, August 1996.

20. T. Tripathi and K. Sivarajan. Computing approximate blocking probabilities in wavelength routed all-optical networks with limited-range wavelength conversion. In *Proceedings of INFOCOM '99*, pages 329–336, March 1999.

21. N. Ghani. Lambda-labeling: A framework for IP-over-WDM using MPLS. *Optical Networks*, 1(2):45–58, April 2000.

22. L. H. Sahasrabuddhe and B. Mukherjee. Light-trees: Optical multicasting for improved performance in wavelength-routed networks. *IEEE Communications*, 37(2):67–73, February 1999.

23. D. Papadimitriou *et al.* Optical multicast in wavelength switched networks – architectural framework. IETF Draft <draft-poj-optical-multicast-01.txt>, July 2001. Work in progress.

24. B. Mukherjee. *Optical Communication Networking*. McGraw-Hill, 1997.

25. V. Sharma and E. A. Varvarigos. Limited wavelength translation in all-optical WDM mesh networks. In *Proceedings of INFOCOM '98*, pages 893–901, March 1999.

26. S. L. Hakimi. Steiner's problem in graphs and its implications. *Networks*, 1:113–133, 1971.

27. Y. Xin, G. N. Rouskas, and H. G. Perros. On the design of MPλS networks. Technical Report TR-01-07, North Carolina State University, Raleigh, NC, July 2001.

28. S. Subramaniam, M. Azizoglu, and A. K. Somani. On the optimal placement of wavelength converters in wavelength-routed networks. In *Proceedings of INFOCOM '98*, pages 902–909, April 1998.

29. M. Ali and J. Deogun. Allocation of splitting nodes in wavelength-routed networks. *Photonic Network Communications*, 2(3):245–263, August 2000.

30. R. Ramaswami and K. N. Sivarajan. Design of logical topologies for wavelength-routed optical networks. *IEEE Journal Selected Areas in Communications*, 14(5):840–851, June 1996.

31. D. Banerjee and B. Mukherjee. A practical approach for routing and wavelength assignment in large wavelength-routed optical networks. *IEEE Journal Selected Areas in Communications*, 14(5):903–908, June 1996.

32. B. Mukherjee *et al.* Some principles for designing a wide-area WDM optical network. *IEEE/ACM Transactions on Networking*, 4(5):684–696, October 1996.

33. Z. Zhang and A. Acampora. A heuristic wavelength assignment algorithm for multihop WDM networks with wavelength routing and wavelength reuse. *IEEE/ACM Transactions on Networking*, 3(3):281–288, June 1995.

34. I. Chlamtac, A. Ganz, and G. Karmi. Lightnets: Topologies for high-speed optical networks. *Journal of Lightwave Technology*, 11:951–961, May/June 1993.
35. S. Banerjee and B. Mukherjee. Algorithms for optimized node placement in shufflenet-based multihop lightwave networks. In *Proceedings of INFOCOM '93*, March 1993.
36. R. Malli, X. Zhang, and C. Qiao. Benefit of multicasting in all-optical networks. In *Proceedings of SPIE*, volume 3531, pages 209–220, November 1998.
37. G. Sahin and M. Azizoglu. Multicast routing and wavelength assignment in wide-area networks. In *Proceedings of SPIE*, volume 3531, pages 196–208, November 1998.
38. E. Lawler. *Combinatorial Optimization: Networks and Matroids*. Holt, Rinehart and Winston, 1976.
39. D. Bertsekas and R. Gallager. *Data Networks*. Prentice Hall, Inc., Englewood Cliffs, NJ, 1992.
40. X. Zhang, J. Y. Wei, and C. Qiao. Constrained multicast routing in WDM networks with sparse light splitting. *Journal of Lightwave Technology*, 18(12):1917–1927, December 2000.
41. J. Strand, A. L. Chiu, and R. Tkach. Issues for routing in the optical layer. *IEEE Communications*, pages 81–96, February 2001.
42. J. Moy. OSPF version 2. RFC 2328, April 1998.
43. A. Chiu et al. Impairments and other constraints on optical layer routing. IETF Draft <draft-ietf-ipo-impairments-00.txt>, May 2001. Work in progress.
44. Y. Zhu, G. N. Rouskas, and H. G. Perros. A comparison of allocation policies in wavelength routing networks. *Photonic Network Communications*, 2(3):265–293, August 2000.
45. Z. Zhang, J. Fu, D. Guo, and L. Zhang. Lightpath routing for intelligent optical networks. *IEEE Network*, 15(4):28–35, July/August 2001.
46. C. Assi, M. Ali, R. Kurtz, and D. Guo. Optical networking and real-time provisioning: An integrated vision for the next-generation internet. *IEEE Network*, 15(4):36–45, July/August 2001.
47. S. Sengupta and R. Ramamurthy. From network design to dynamic provisioning and restoration in optical cross-connect mesh networks: An architectural and algorithmic overview. *IEEE Network*, 15(4):46–54, July/August 2001.
48. The internet engineering task force. http://www.ietf.org.
49. Optical domain service interconnect. http://www.odsi-coalition.com.
50. The optical internetworking forum. http://www.oiforum.com.
51. G. Bernstein, R. Coltun, J. Moy, A. Sodder, and K. Arvind. ODSI functional specification version 1.4. ODSI Coalition, August 2000.
52. User network interface (UNI) 1.0 signaling specification. OIF2000.125.6, September 2001.
53. B. Rajagopalan et al. IP over optical networks – a framework. IETF Draft <draft-many-ip-optical-framework-03.txt>, March 2001. Work in progress.
54. J. P. Lang et al. Link management protocol (LMP). IETF Draft <draft-ietf-mpls-lmp-02.txt>, September 2001. Work in progress.
55. K. Kompella et al. OSPF extensions in support of generalized MPLS. IETF Draft <draft-ietf-ccamp-ospf-gmpls-extensions-00.txt>, September 2001. Work in progress.
56. D. Katz, D. Yeung, and K. Kompella. Traffic engineering extensions to OSPF. IETF Draft <draft-katz-yeung-ospf-traffic-06.txt>, October 2001. Work in progress.

57. D. Awduche *et al.* RSVP-TE: Extensions to RSVP for LSP tunnels. IETF Draft <draft-ietf-mpls-rsvp-lsp-tunnel-08.txt>, February 2001. Work in progress.

58. O. Aboul-Magd *et al.* Constraint-based LSP setup using LDP. IETF Draft <draft-ietf-mpls-cr-ldp-05.txt>, February 2001. Work in progress.

59. R. Braden *et al.* Resource reservation protocol – version 1. RFC 2205, September 1997.

60. L. Andersson, P. Doolan, N. Feldman, A. Fredette, and B. Thomas. LDP specification. RFC 3036, January 2001.

61. P. Ashwood-Smith *et al.* Generalized MPLS signaling – RSVP-TE extensions. IETF Draft <draft-ietf-mpls-generalized-rsvp-te-05.txt>, October 2001. Work in progress.

62. P. Ashwood-Smith *et al.* Generalized MPLS signaling – CR-LDP extensions. IETF Draft <draft-ietf-mpls-generalized-cr-ldp-04.txt>, July 2001. Work in progress.

63. B. Rajagopalan, J. Luciani, D. Awduche, B. Cain, B. Jamoussi, and D. Saha. IP over optical networks: A framework. IETF Draft <draft-ietf-ipo-framework-01.txt>, February 2002. Work in progress.

64. H. Perros. *An Introduction to ATM Networks.* Wiley, 2001.

65. D. Durham (Ed.), J. Boyle, R. Cohen, S. Herzog, R. Rajan, and A. Sastry. The COPS (common open policy service) protocol. RFC 2748, January 2000.

66. S. Herzog (Ed.), J. Boyle, R. Cohen, D. Durham, R. Rajan, and A. Sastry. COPS usage for RSVP. RFC 2749, January 2000.

67. S. Yao, S. Dixit, and B. Mukherjee. Advances in photonic packet switching: An overview. *IEEE Communications*, 38(2):84–94, February 2000.

68. H. Perros. *Queueing Networks with Blocking: Exact and Approximate Solutions.* Oxford University Press, 1994.

69. L. Xu, H. G. Perros, and G. N. Rouskas. Techniques for optical packet switching and optical burst switching. *IEEE Communications*, 39(1):136–142, January 2001.

70. S. L. Danielsen *et al.* Analysis of a WDM packet switch with improved performance under bursty traffic conditions due to tunable wavelength converters. *IEEE/OSA Journal of Lightwave Technology*, 16(5):729–735, May 1998.

71. I. Baldine, G. N. Rouskas, H. G. Perros, and D. Stevenson. JumpStart: A just-in-time signaling architecture for WDM burst-switched networks. *IEEE Communications*, 40(2):82–89, February 2002.

72. C. Qiao and M. Yoo. Optical burst switching (OBS)-A new paradigm for an optical Internet. *Journal of High Speed Networks*, 8(1):69–84, January 1999.

73. J. Y. Wei and R. I. McFarland. Just-in-time signaling for WDM optical burst switching networks. *Journal of Lightwave Technology*, 18(12):2019–2037, December 2000.

Author Index

Basagni, Stefano   101

Capra, Licia   20
Crowcroft, Jon   1

Davide, Fabrizio   124
Deri, Luca   83

Emmerich, Wolfgang   20

Gantenbein, Dieter   83

Loreti, Pierpaolo   124

Lunghi, Massimiliano   124

Mascolo, Cecilia   20
Molva, Refik   59

Pannetrat, Alain   59
Perros, Harry G.   155
Pratt, Ian   1

Riva, Giuseppe   124
Rouskas, George N.   155

Vatalaro, Francesco   124

Lecture Notes in Computer Science

For information about Vols. 1–2446

please contact your bookseller or Springer-Verlag

Vol. 2447: D.J. Hand, N.M. Adams, R.J. Bolton (Eds.), Pattern Detection and Discovery. Proceedings, 2002. XII, 227 pages. 2002. (Subseries LNAI).

Vol. 2448: P. Sojka, I. Kopecýek, K. Pala (Eds.), Text, Speech and Dialogue. Proceedings, 2002. XII, 481 pages. 2002. (Subseries LNAI).

Vol. 2449: L. Van Gool (Ed.), Pattern Recognotion. Proceedings, 2002. XVI, 628 pages. 2002.

Vol. 2451: B. Hochet, A.J. Acosta, M.J. Bellido (Eds.), Integrated Circuit Design. Proceedings, 2002. XVI, 496 pages. 2002.

Vol. 2452: R. Guigó, D. Gusfield (Eds.), Algorithms in Bioinformatics. Proceedings, 2002. X, 554 pages. 2002.

Vol. 2453: A. Hameurlain, R. Cicchetti, R. Traunmüller (Eds.), Database and Expert Systems Applications. Proceedings, 2002. XVIII, 951 pages. 2002.

Vol. 2454: Y. Kambayashi, W. Winiwarter, M. Arikawa (Eds.), Data Warehousing and Knowledge Discovery. Proceedings, 2002. XIII, 339 pages. 2002.

Vol. 2455: K. Bauknecht, A M. Tjoa, G. Quirchmayr (Eds.), E-Commerce and Web Technologies. Proceedings, 2002. XIV, 414 pages. 2002.

Vol. 2456: R. Traunmüller, K. Lenk (Eds.), Electronic Government. Proceedings, 2002. XIII, 486 pages. 2002.

Vol. 2457: T. Yakhno (Ed.), Advances in Information Systems. Proceedings, 2002. XII, 436 pages. 2002.

Vol. 2458: M. Agosti, C. Thanos (Eds.), Research and Advanced Technology for Digital Libraries. Proceedings, 2002. XVI, 664 pages. 2002.

Vol. 2459: M.C. Calzarossa, S. Tucci (Eds.), Performance Evaluation of Complex Systems: Techniques and Tools. Proceedings, 2002. VIII, 501 pages. 2002.

Vol. 2460: J.-M. Jézéquel, H. Hussmann, S. Cook (Eds.), ÇUMLÈ 2002 – The Unified Modeling Language. Proceedings, 2002. XII, 449 pages. 2002.

Vol. 2461: R. Möhring, R. Raman (Eds.), Algorithms – ESA 2002. Proceedings, 2002. XIV, 917 pages. 2002.

Vol. 2462: K. Jansen, S. Leonardi, V. Vazirani (Eds.), Approximation Algorithms for Combinatorial Optimization. Proceedings, 2002. VIII, 271 pages. 2002.

Vol. 2463: M. Dorigo, G. Di Caro, M. Sampels (Eds.), Ant Algorithms. Proceedings, 2002. XIII, 305 pages. 2002.

Vol. 2464: M. O'Neill, R.F.E. Sutcliffe, C. Ryan, M. Eaton, N. Griffith (Eds.), Artificial Intelligence and Cognitive Science. Proceedings, 2002. XI, 247 pages. 2002. (Subseries LNAI).

Vol. 2465: H. Arisawa, Y. Kambayashi, V. Kumar, H.C. Mayr, I. Hunt (Eds.), Conceptual Modeling for New Information Systems Technologies. Proceedings, 2001. XVII, 500 pages. 2002.

Vol. 2466: M. Beetz, J. Hertzberg, M. Ghallab, M.E. Pollack (Eds.), Advances in Plan-Based Control of Robotic Agents. Proceedings, 2001. VIII, 291 pages. 2002. (Subseries LNAI).

Vol. 2467: B. Christianson, B. Crispo, J.A. Malcolm, M. Roe (Eds.), Security Protocols. Proceedings, 2001. IX, 241 pages. 2002.

Vol. 2469: W. Damm, E.-R. Olderog (Eds.), Formal Techniques in Real-Time and Fault-Tolerant Systems. Proceedings, 2002. X, 455 pages. 2002.

Vol. 2470: P. Van Hentenryck (Ed.), Principles and Practice of Constraint Programming – CP 2002. Proceedings, 2002. XVI, 794 pages. 2002.

Vol. 2471: J. Bradfield (Ed.), Computer Science Logic. Proceedings, 2002. XII, 613 pages. 2002.

Vol. 2473: A. Gomez-Perez, V.R. Benjamins, Knowledge Engineering and Knowledge Management. Proceedings, 2002. XI, 402 pages. 2002. (Subseries LNAI).

Vol. 2474: D. Kranzlmüller, P. Kacsuk, J. Dongarra, J. Volkert (Eds.), Recent Advances in Parallel Virtual Machine and Message Passing Interface. Proceedings, 2002. XVI, 462 pages. 2002.

Vol. 2475: J.J. Alpigini, J.F. Peters, A. Skowron, N. Zhong (Eds.), Rough Sets and Current Trends in Computing. Proceedings, 2002. XV, 640 pages. 2002. (Subseries LNAI).

Vol. 2476: A.H.F. Laender, A.L. Oliveira (Eds.), String Processing and Information Retrieval. Proceedings, 2002. XI, 337 pages. 2002.

Vol. 2477: M.V. Hermenegildo, G. Puebla (Eds.), Static Analysis. Proceedings, 2002. XI, 527 pages. 2002.

Vol. 2478: M.J. Egenhofer, D.M. Mark (Eds.), Geographic Information Science. Proceedings, 2002. X, 363 pages. 2002.

Vol. 2479: M. Jarke, J. Koehler, G. Lakemeyer (Eds.), KI 2002: Advances in Artificial Intelligence. Proceedings, 2002. XIII, 327 pages. (Subseries LNAI).

Vol. 2480: Y. Han, S. Tai, D. Wikarski (Eds.), Engineering and Deployment of Cooperative Information Systems. Proceedings, 2002. XIII, 564 pages. 2002.

Vol. 2483: J.D.P. Rolim, S. Vadhan (Eds.), Randomization and Approximation Techniques in Computer Science. Proceedings, 2002. VIII, 275 pages. 2002.

Vol. 2484: P. Adriaans, H. Fernau, M. van Zaanen (Eds.), Grammatical Inference: Algorithms and Applications. Proceedings, 2002. IX, 315 pages. 2002. (Subseries LNAI).

Vol. 2485: A. Bondavalli, P. Thevenod-Fosse (Eds.), Dependable Computing EDCC-4. Proceedings, 2002. XIII, 283 pages. 2002.

Vol. 2486: M. Marinaro, R. Tagliaferri (Eds.), Neural Nets. Proceedings, 2002. IX, 253 pages. 2002.

Vol. 2487: D. Batory, C. Consel, W. Taha (Eds.), Generative Programming and Component Engineering. Proceedings, 2002. VIII, 335 pages. 2002.

Vol. 2488: T. Dohi, R. Kikinis (Eds), Medical Image Computing and Computer-Assisted Intervention – MICCAI 2002. Proceedings, Part I. XXIX, 807 pages. 2002.

Vol. 2489: T. Dohi, R. Kikinis (Eds), Medical Image Computing and Computer-Assisted Intervention – MICCAI 2002. Proceedings, Part II. XXIX, 693 pages. 2002.

Vol. 2490: A.B. Chaudhri, R. Unland, C. Djeraba, W. Lindner (Eds.), XML-Based Data Management and Multimedia Engineering – EDBT 2002. Proceedings, 2002. XII, 652 pages. 2002.

Vol. 2491: A. Sangiovanni-Vincentelli, J. Sifakis (Eds.), Embedded Software. Proceedings, 2002. IX, 423 pages. 2002.

Vol. 2492: F.J. Perales, E.R. Hancock (Eds.), Articulated Motion and Deformable Objects. Proceedings, 2002. X, 257 pages. 2002.

Vol. 2493: S. Bandini, B. Chopard, M. Tomassini (Eds.), Cellular Automata. Proceedings, 2002. XI, 369 pages. 2002.

Vol. 2495: C. George, H. Miao (Eds.), Formal Methods and Software Engineering. Proceedings, 2002. XI, 626 pages. 2002.

Vol. 2496: K.C. Almeroth, M. Hasan (Eds.), Management of Multimedia in the Internet. Proceedings, 2002. XI, 355 pages. 2002.

Vol. 2497: E. Gregori, G. Anastasi, S. Basagni (Eds.), Advanced Lectures on Networking. XI, 195 pages. 2002.

Vol. 2498: G. Borriello, L.E. Holmquist (Eds.), UbiComp 2002: Ubiquitous Computing. Proceedings, 2002. XV, 380 pages. 2002.

Vol. 2499: S.D. Richardson (Ed.), Machine Translation: From Research to Real Users. Proceedings, 2002. XXI, 254 pages. 2002. (Subseries LNAI).

Vol. 2501: D. Zheng (Ed.), Advances in Cryptology – ASIACRYPT 2002. Proceedings, 2002. XIII, 578 pages. 2002.

Vol. 2502: D. Gollmann, G. Karjoth, M. Waidner (Eds.), Computer Security – ESORICS 2002. Proceedings, 2002. X, 281 pages. 2002.

Vol. 2503: S. Spaccapietra, S.T. March, Y. Kambayashi (Eds.), Conceptual Modeling – ER 2002. Proceedings, 2002. XX, 480 pages. 2002.

Vol. 2504: M.T. Escrig, F. Toledo, E. Golobardes (Eds.), Topics in Artificial Intelligence. Proceedings 2002. XI, 432 pages. 2002. (Subseries LNAI).

Vol. 2506: M. Feridun, P. Kropf, G. Babin (Eds.), Management Technologies for E-Commerce and E-Business Applications. Proceedings, 2002. IX, 209 pages. 2002.

Vol. 2507: G. Bittencourt, G.L. Ramalho (Eds.), Advances in Artificial Intelligence. Proceedings, 2002. XIII, 418 pages. 2002. (Subseries LNAI).

Vol. 2508: D. Malkhi (Ed.), Distributed Computing. Proceedings, 2002. X, 371 pages. 2002.

Vol. 2509: C.S. Calude, M.J. Dinneen, F. Peper (Eds.), Unconventional Models in Computation. Proceedings, 2002. VIII, 331 pages. 2002.

Vol. 2510: H. Shafazand, A Min Tjoa (Eds.), EurAsia-ICT 2002: Information and Communication Technology. Proceedings, 2002. XXIII, 1020 pages. 2002.

Vol. 2511: B. Stiller, M. Smirnow, M. Karsten, P. Reichl (Eds.), From QoS Provisioning to QoS Charging. Proceedings, 2002. XIV, 348 pages. 2002.

Vol. 2513: R. Deng, S. Qing, F. Bao, J. Zhou (Eds.), Information and Communications Security. Proceedings, 2002. XII, 496 pages. 2002.

Vol. 2514: M. Baaz, A. Voronkov (Eds.), Logic for Programming, Artificial Intelligence, and Reasoning. Proceedings 2002. XIII, 465 pages. 2002. (Subseries LNAI).

Vol. 2515: F. Boavida, E. Monteiro, J. Orvalho (Eds.), Protocols and Systems for Interactive Distributed Multimedia. Proceedings, 2002. XIV, 372 pages. 2002.

Vol. 2516: A. Wespi, G. Vigna, L. Deri (Eds.), Recent Advances in Intrusion Detection. Proceedings, 2002. X, 327 pages. 2002.

Vol. 2517: M.D. Aagaard, J.W. O'Leary (Eds.), Formal Methods in Computer-Aided Design. Proceedings, 2002. XI, 399 pages. 2002.

Vol. 2518: P. Bose, P. Morin (Eds.), Algorithms and Computation. Proceedings, 2002. XIII, 656 pages. 2002.

Vol. 2519: R. Meersman, Z. Tari, et al. (Eds.), On the Move to Meaningful Internet Systems 2002: CoopIS, DOA, and ODBASE. Proceedings, 2002. XXIII, 1367 pages. 2002.

Vol. 2521: A. Karmouch, T. Magedanz, J. Delgado (Eds.), Mobile Agents for Telecommunication Applications. Proceedings 2002. XII, 317 pages. 2002.

Vol. 2522: T. Andreasen, A. Motro, H. Christiansen, H. Legind Larsen (Eds.), Flexible Query Answering. Proceedings 2002. XI, 386 pages. 2002. (Subseries LNAI).

Vol. 2525: H.H. Bülthoff, S.-Whan Lee, T.A. Poggio, C. Wallraven (Eds.), Biologically Motivated Computer Vision. Proceedings 2002. XIV, 662 pages. 2002.

Vol. 2526: A. Colosimo, A. Giuliani, P. Sirabella (Eds.), Medical Data Analysis. Proceedings 2002. IX, 222 pages. 2002.

Vol. 2527: F.J. Garijo, J.C. Riquelme, M. Toro (Eds.), Advances in Artificial Intelligence – IBERAMIA 2002. Proceedings 2002. XVIII, 955 pages. 2002. (Subseries LNAI).

Vol. 2528: M.T. Goodrich, S.G. Kobourov (Eds.), Graph Drawing. Proceedings 2002. XIII, 384 pages. 2002.

Vol. 2529: D.A. Peled, M.Y. Vardi (Eds.), Formal Techniques for Networked and Distributed Sytems – FORTE 2002. Proceedings 2002. XI, 371 pages. 2002.

Vol. 2534: S. Lange, K. Satoh, C.H. Smith (Ed.), Discovery Science. Proceedings 2002. XIII, 464 pages. 2002.

Vol. 2535: N. Suri (Ed.), Mobile Agents. Proceedings 2002. X, 203 pages. 2002.

Vol. 2536: M. Parashar (Ed.), Grid Computing – GRID 2002. Proceedings 2002. XI, 318 pages. 2002.

Vol. 2540: W.I. Grosky, F. Plášil (Eds.), SOFSEM 2002: Theory and Practice of Informatics. Proceedings 2002. X, 289 pages. 2002.